ENERGY 2023

The Thirteenth International Conference on Smart Grids, Green Communications and IT Energy-aware Technologies

March 13th - 17th, 2023

Barcelona, Spain

ENERGY 2023 Editors

Eric MSP Veith, OFFIS e.V. — Institute for Information Technology, R&D Division Energy - Oldenburg, Germany

Printed from e-media with permission by:

Curran Associates, Inc.
57 Morehouse Lane
Red Hook, NY 12571

Some format issues inherent in the e-media version may also appear in this print version.

Printed with permission by Curran Associates, Inc. (2023)

International Academy, Research, and Industry Association (IARIA)
412 Derby Way
Wilmington, DE 19810

Phone: (408) 893-6407
Fax: (408) 527-6351

petre@iaria.org

Additional copies of this publication are available from:

Curran Associates, Inc.
57 Morehouse Lane
Red Hook, NY 12571 USA
Phone: 845-758-0400
Fax: 845-758-2633
Email: curran@proceedings.com
Web: www.proceedings.com

Table of Contents

Programmable Logic Controllers - Insecure by Design? A Survey 1
Benedikt Geisler and Markus Kucera

PIV of Flow Convection Induced by Solar Sphere to Generate Power 5
Hassan Abdulmouti

An Architecture for Reliable Learning Agents in Power Grids 13
Eric Veith

Development of Occupants' Behavior Model in Urban Scale Using Dynamic Time Warping and Particle Swarm Optimization Algorithms Based on National Lifetime Survey 17
Hengxuan Wang and Sumiyoshi Daisuke

Machine Learning and Optimisation to Improve Energy Utilisation Efficiency 23
Srinath Ramagiri, Evelyne El Masri, Arman Zonuzi, Ahmed Teyeb, Shehan Lowe, and Tat-Hean Gan

An Islanded Community Solar Microgrid with Capability of Future Fractal Growth 29
Mark Apperley and Tama Toki

The Implementation of the Automatic Dispatching System (ADS) to Support the Smart Grid Pilot Project for Distribution Grid Improvement in Sumba Island 36
Dimas Fiddiansyah and Hanny Benchmark

Implementation of a Fully-automated Optimized Fog-computing-based IoT-controlled PV Network 42
Adel AlWaqfi, Mohammad Abdel-Rahman, Yunis Al-Qreenawi, Zain AlHwaidy, and Ahmad Abu-El-Haija

Spatial Analysis to Identify the Need for Additional EV Charging 49
Vivian Sultan, Benjamin Pezzillo, Roxana Cuevas, Solange Ruiz, Jinhua Tang, and Amanda Acosta

Business Model Considerations for a Solution to Optimize and Diagnose Solar Panel Installations 57
Lasse Berntzen, Saeed Teimourzadeh, Paula Anghelita, Qian Meng, and Viorel Ursu

Performance Analysis of Single and Multi-step Short-term Load Forecasts Using Multilayer Perceptron 63
Athanasios Ioannis Arvanitidis, Dimitrios Kontogiannis, Georgios Vontzos, Vasileios Laitsos, Dimitrios Bargiotas, and Miltiadis Alamaniotis

ENERGY 2023

Foreword

The Thirteenth International Conference on Smart Grids, Green Communications and IT Energy-aware Technologies (ENERGY 2023), held between March 13 – 17, 2023, continued the event considering Green approaches for Smart Grids and IT-aware technologies. It addressed fundamentals, technologies, hardware and software needed support, and applications and challenges.

There is a perceived need for a fundamental transformation in IP communications, energy-aware technologies and the way all energy sources are integrated. This is accelerated by the complexity of smart devices, the need for special interfaces for an easy and remote access, and the new achievements in energy production. Smart Grid technologies promote ways to enhance efficiency and reliability of the electric grid, while addressing increasing demand and incorporating more renewable and distributed electricity generation. The adoption of data centers, penetration of new energy resources, large dissemination of smart sensing and control devices, including smart home, and new vehicular energy approaches demand a new position for distributed communications, energy storage, and integration of various sources of energy.

We take here the opportunity to warmly thank all the members of the ENERGY 2023 Technical Program Committee, as well as the numerous reviewers. The creation of such a high quality conference program would not have been possible without their involvement. We also kindly thank all the authors who dedicated much of their time and efforts to contribute to ENERGY 2023. We truly believe that, thanks to all these efforts, the final conference program consisted of top quality contributions.

Also, this event could not have been a reality without the support of many individuals, organizations, and sponsors. We are grateful to the members of the ENERGY 2023 organizing committee for their help in handling the logistics and for their work to make this professional meeting a success.

We hope that ENERGY 2023 was a successful international forum for the exchange of ideas and results between academia and industry and for the promotion of progress in the fields of smart grids, green communications and IT energy-aware technologies.

We are convinced that the participants found the event useful and communications very open. We also hope that Barcelona provided a pleasant environment during the conference and everyone saved some time for exploring this beautiful city.

ENERGY 2023 Chairs:

ENERGY 2023

Committee

ENERGY 2023 Steering Committee

Eric MSP Veith, OFFIS e.V. – Oldenburg, Germany
Dragan Obradovic, Siemens - Corporate Technology, Munich, Germany
Mark Apperley, University of Waikato, New Zealand
Michael Negnevitsky, University of Tasmania, Australia
Vivian Sultan, California State University Los Angeles, USA
Steffen Fries, Siemens, Germany

ENERGY 2023 Publicity Chairs

Sandra Viciano Tudela, Universitat Politecnica de Valencia, Spain
José Miguel Jiménez, Universitat Politecnica de Valencia, Spain

ENERGY 2023 Technical Program Committee

Santosh Aditham, University of South Florida, USA
Kodjo Agbossou, Université du Québec à Trois-Rivières, Canada
Awadelrahman M. A. Ahmed, University of Oslo, Norway
Miltos Alamaniotis, University of Texas at San Antonio, USA
Kamal Al-Haddad, École de technologie supérieure, Montreal, Canada
Ahmed Al-Salaymeh, The University of Jordan, Amman, Jordan
Amjad Anvari-Moghaddam, AAU Energy, Denmark
Mark Apperley, University of Waikato, New Zealand
Parimal Acharjee, NIT Durgapur, India
Paranietharan Arunagirinathan, Clemson University, USA
Ashrant Aryal, Texas A&M University, USA
Adela Bara, Bucharest University of Economic Studies, Department of Economic Informatics and Cybernetics, Bucharest, Romania
Navid Bayati, Aalborg University, Denmark
Andreas Berl, Technische Hochschule Deggendorf, Germany
Rico Berner, Humboldt-Universität zu Berlin | Institut für Physik, Germany
Lasse Berntzen, University of South-Eastern Norway, Norway
Vito Calderaro, University of Salerno, Italy
M. Girish Chandra, TCS Research & Innovation, India
Dana-Alexandra Ciupageanu, University Politehnica of Bucharest, Romania
Daniele Codetta, University of Piemonte Orientale, Italy
Luigi Costanzo, Università degli Studi della Campania Luigi Vanvitelli, Italy
Fabio D'Agostino, University of Genova, Italy
Thusitha Dayaratne, Monash University, Australia
Payman Dehghanian, The George Washington University, USA
Margot Deruyck, Universiteit Gent - IMEC - WAVES, Belgium
Giovanna Dondossola, RSE, Italy
Virgil Dumbrava, University POLITEHNICA of Bucharest, Romania

Kevin Ellett, Indiana University, USA
Tatiana Endrjukaite, Transport and Telecommunication Institute, Riga, Latvia
Meisam Farrokhifar, Eindhoven University of Technology, The Netherlands
Sébastien Faye, Luxembourg Institute of Science and Technology (LIST), Luxembourg
Wendy Flores-Fuentes, Autonomous University of Baja California, Mexicali, Mexico
Mahmoud Fotuhi-Firuzabad, Sharif University of Technology, Tehran, Iran
Steffen Fries, Siemens, Germany
Vincenzo Galdi, University of Salerno, Italy
Francisco M. Gonzalez-Longatt, University of South-Eastern Norway, Norway
Saman K. Halgamuge, The University of Melbourne, Australia
Yunzhi Huang, Pacific Northwest National Laboratory - U.S. Department of Energy, USA
Md. Minarul Islam, University of Dhaka, Bangladesh
Philip Johnson, University of Hawaii, USA
Michael Kuhn, Otto von Guericke University Magdeburg, Germany
Rajat Kumar, Dhirubhai Ambani Institute of information and communication technology, Gandhinagar, India
Tobias Küster, Technische Universität Berlin (TU Berlin), Germany
Sebastian Lawrenz, Clausthal University of Technology, Germany
Duc Van Le, Nanyang Technological University, Singapore
Gerard Ledwich, Queensland University of Technology, Australia
Yiu-Wing Leung, Hong Kong Baptist University, Kowloon Tong, Hong Kong
Qinghua Li, University of Arkansas, USA
Li Hong Idris Lim, University of Glasgow, UK
Zhenhua Liu, Stony Brook University (SUNY at Stony Brook), USA
Pascal Maussion, Toulouse INP / CNRS LAPLACE, France
Amin Mokari, School of Electrical Engineering & Robotics | Queensland University of Technology, Australia
Hugo Morais, Universidade de Lisboa, Portugal
Fabio Mottola, University of Naples Federico II, Italy
Emmanuel Mudaheranwa, Cardiff University, UK / Rwanda Polytechnic, Rwanda
Gero Mühl, Universitaet Rostock, Germany
Hamidreza Nazaripouya, University of California, Riverside, USA
Michael Negnevitsky, University of Tasmania, Australia
Simona Olmi, Consiglio Nazionale delle Ricerche - Istituto dei Sistemi Complessi, Italy
Claudiu Oprea, Technical University of Cluj-Napoca, Romania
Youssef Ounejjar, ETS, Montreal, Canada
Shalini Pandey, University of Minnesota, USA
Sanjeev Pannala, Washington State University, USA
Thanasis Papaioannou, Athens University of Economics and Business (AUEB), Greece
Marco Pasetti, University of Brescia, Italy
Nilavra Pathak, University of Maryland Baltimore County, USA
Lawrence Pileggi, Carnegie Mellon University, USA
Marco Pruckner, Friedrich-Alexander-University Erlangen-Nürnberg, Germany
Venkata Ramakrishna P., Tata Consultancy Services, India
Djamila Rekioua, University of Bejaia, Algeria
Huamin Ren, Kristiania University College, Oslo, Norway
Jan Richling, South Westphalia University of Applied Sciences, Germany
Stefano Rinaldi, University of Brescia, Italy

Copyright Information

For your reference, this is the text governing the copyright release for material published by IARIA.

The copyright release is a transfer of publication rights, which allows IARIA and its partners to drive the dissemination of the published material. This allows IARIA to give articles increased visibility via distribution, inclusion in libraries, and arrangements for submission to indexes.

I, the undersigned, declare that the article is original, and that I represent the authors of this article in the copyright release matters. If this work has been done as work-for-hire, I have obtained all necessary clearances to execute a copyright release. I hereby irrevocably transfer exclusive copyright for this material to IARIA. I give IARIA permission or reproduce the work in any media format such as, but not limited to, print, digital, or electronic. I give IARIA permission to distribute the materials without restriction to any institutions or individuals. I give IARIA permission to submit the work for inclusion in article repositories as IARIA sees fit.

I, the undersigned, declare that to the best of my knowledge, the article is does not contain libelous or otherwise unlawful contents or invading the right of privacy or infringing on a proprietary right.

Following the copyright release, any circulated version of the article must bear the copyright notice and any header and footer information that IARIA applies to the published article.

IARIA grants royalty-free permission to the authors to disseminate the work, under the above provisions, for any academic, commercial, or industrial use. IARIA grants royalty-free permission to any individuals or institutions to make the article available electronically, online, or in print.

IARIA acknowledges that rights to any algorithm, process, procedure, apparatus, or articles of manufacture remain with the authors and their employers.

I, the undersigned, understand that IARIA will not be liable, in contract, tort (including, without limitation, negligence), pre-contract or other representations (other than fraudulent misrepresentations) or otherwise in connection with the publication of my work.

Exception to the above is made for work-for-hire performed while employed by the government. In that case, copyright to the material remains with the said government. The rightful owners (authors and government entity) grant unlimited and unrestricted permission to IARIA, IARIA's contractors, and IARIA's partners to further distribute the work.

Programmable Logic Controllers - Insecure by Design? A Survey.

Benedikt Geisler
Faculty of Computer Science and Mathematics
Ostbayerische Technische Hochschule Regensburg
Regensburg, Germany
email: benedikt1.geisler@st.oth-regensburg.de

Markus Kucera
Faculty of Computer Science and Mathematics
Ostbayerische Technische Hochschule Regensburg
Regensburg, Germany
email: markus.kucera@oth-regensburg.de

Abstract—**Any cyber-physical system, including critical infrastructure such as a smart grid, is very likely being controlled by an Industrial Control System (ICS). However, ICSs have long been neglected in terms of security mechanisms. This work presents an overview of the current situation by conducting a literature review, focusing on attack vectors against Programmable Logic Controllers (PLC). Due to proprietary protocols and operating systems, it will present attacks against four major vendors: Siemens, Allen Bradley, Schneider, and Beckhoff.**

Index Terms—**OT, PLC, ICS, Modbus, Profinet**

I. Introduction

Stuxnet [1], destroying almost 1000 centrifuges in the Iranian uranium enrichment facility in Natanz, was the first attack against an Industrial Control System (ICS) that gained broad attention and raised security awareness in the area of Operation Technology (OT). Another attack targeting ICSs caused a power outage in Ukraine which affected 225.000 people [2]. More, often sophisticated, attacks have been carried out since [3].

Predominantly these attacks target PLCs. These differ in several important aspects from traditional computers: (i) They do not interact with data, but with the physical world instead. (ii) Thus, their main interest is not confidentiality, but reliably running a continuous process. (iii) Their lifespan is commonly between 15 and 20 years. (iv) Once installed, they hardly ever get patched. (v) They use proprietary firmware or operating systems (OS). (vi) They execute their programs in continuous, real-time cycles. Thus, due to their different primary objectives, design, and use they have to be treated differently [4].

In this work, we present an overview of known attack vectors against ICSs and show the underlying common security weaknesses (Section III). Secure coding practices and guidelines exist, but they are not the focus of this work. We touch on them only briefly in Section IV.

II. Taxonomy

There are many ways to structure ICS security [5][6][7]. We will use a target-based structure in our work for two reasons: (i) most publications presenting attacks follow this approach and (ii) due to proprietary protocols and operating systems these attacks are most often vendor specific. However, this approach can be seen as *pars pro toto* since many of the principles presented can be transferred from one vendor to another.

III. Attacks

In the past, an *air gap* (not connecting the OT and Information Technology (IT) networks) was perceived as adequate protection against attacks. However, with the advent of Manufacturing Execution Systems, remote access, and the Industrial Internet of Things this no longer holds. Next, Security by Obscurity (use of closed-source, vendor-specific protocols and security mechanisms) has numerously been compromised as we will show next.

All presented attacks assume that the adversary has already gained access to the IT network and can move laterally to the OT network, as has been the case in past attacks.

A. Siemens S7

After the Stuxnet attack in 2011, Beresford [8] was the first to exploit vulnerabilities of Siemens S7-300 and S7-1200. His work shows in detail how to gain access to a PLC by first capturing session data and then replaying it to the PLC. This can further be extended to altering the control logic or disabling the authentication mechanisms altogether without having access to the engineering workstation. These attacks are possible due to the use of insecure protocols in ICSs.

Following up on the replay attack, [9] extend this procedure by reverse-engineering the password encoding scheme. They succeed, revealing the custom eight-byte XOR encoding scheme. This allows not only to update the password of the PLC but also to clear arbitrary PLC memory which effectively renders the PLC useless.

As [10] remarks, the password can also be revealed by using an exhaustive search due to the small key space of only eight bits.

By implementing a PLC worm [11], PLCs can infect one another, having the worm automatically propagate through the whole OT network segment.

However, this worm can be detected since TIA Portal engineering software can be used to retrieve the code from the PLC. Building upon existing reverse engineering findings, [12] show that it is possible to disguise the code change. The source code in the PLC exists in a *source object*, but when communicating with the PLC a *run object* is sent. This can be modified to a custom behavior, resulting in a different program being run in the PLC than shown on the engineering station.

B. Allen Bradley

Attacks against PLCs require data transmission via the network. This makes Network Intrusion Detection Systems (NIDS), such as Anagram [13], a natural countermeasure. However, it is possible to develop stealthy attacks by either modifying the signature of the packet header (*Data Execution Attack*) or by fragmentation of the data with added noise padding (*Fragmentation and Noise Padding*). Both attacks are successfully carried out in [14] without being detected by a NIDS and attacked a Schneider Modicon M221, as well as an Allen Bradley MicroLogix 1400.

Vendor-specific engineering software is used to send and retrieve compiled logic to and from the PLC. Thus, it can also be used for forensics in case of control logic injection attacks. However, as [15] show, this is no longer the case if a *Denial of Engineering Operations* attack is used. They show three different versions: (i) Hiding infected ladder logic from the engineering software, (ii) crashing the engineering software upon retrieving code from the PLC, and (iii) injecting a crafted ladder logic program to the PLC that crashes the engineering software. While the former two are *man-in-the-middle-attacks*, the latter is the stealthiest since it allows the attacker to leave the network after the attack. To detect these attacks, the authors also developed a decompiler for ladder logic that can completely restore the ladder program from network traffic and thus makes it possible to detect the injected control logic.

C. Schneider

The Schneider Tricon PLC employs Triple Modular Redundancy, using *two-out-of-three voting*. It is widely used in nuclear power plants. The software is downloaded simultaneously to all three processors, making this PLC susceptible to common mode failures induced by software, such as a cyber attack. Two attacks are proposed by [16], namely *latent attack* which downloads valid but incorrect control logic to the PLC, and *immediate failure attack* which transfers invalid data to the PLC, leading to a denial of engineering and an error on the PLC. While the first causes an incorrect behavior of the PLC and at the same time deceives the operator, the latter leads to a major downtime of the whole system since a complete reset and new program download becomes necessary.

A sophisticated attack against a Schneider Modicon M221 is shown in [17]. The authors propose the fully automated attack tool *CLIK* that consists of four stages: (i) stealing control logic binary from the PLC, (ii) decompiling the stolen binary to source code, (iii) infecting the control logic in the PLC, and (iv) concealment of infection from engineering software using a virtual PLC.

D. Beckhoff

With the use of common operating systems such as Windows CE or Windows 10, Beckhoff differs from other vendors who all use proprietary OS. However, this also makes the PLC susceptible to attacks known from the IT world. Bonney et al. [18] examine a CX5020 PLC and find several possible attack vectors due to plaintext transmitted connection setups (including user name and password), by default enabled webserver, and insecure default user name and password for Virtual Private Network.

E. Modbus

Modbus is a popular, vendor-agnostic protocol used in ICSs. Attacks against SCADA systems can be carried out using this protocol, as [19] show. The authors identify four attack classes: reconnaissance, response and measurement injection, command injection, and denial of service. For each class, they present several concrete attacks.

F. Open Platform Communications Unified Architecture

Open Platform Communications Unified Architecture (OPC UA) is a platform-independent service-oriented architecture that is widely used in the industry and supported by all major vendors. It is mainly used for non-real-time data exchange between PLCs and a variety of clients but with OPC UA over Time-Sensitive-Networking (TSN), it can also be used for real-time communication. OPC UA libraries exist for all major programming languages. OPC UA has been designed with a strong focus on security by integrating the following mechanisms: user security by using a user security token, application security by using digitally signed X.509 certificates, and transport-level security by signing and encrypting each message [20]. A report of the German Federal Office for Information Security (BSI) attests adequate protection against numerous threats, while *denial of service* and *server profiling* can only be reduced by its protection mechanisms [21]. It also notes that no systematic errors could be detected. However, as [22] note, provisioning has not been examined. They show that trust on first use (TOFU) is used for provisioning, thus undermining the security guarantees of OPC UA if the adversary gains access during this first phase. They also note that provisioning is both overly complex and often vaguely documented, leading to misconfiguration or disabling security features.

OPC UA security can also be weakened by major security flaws in its artifacts, as [23] show. The authors examine 48 OPC UA-enabled artifacts, both products from vendors and open-source libraries. Their main findings include disabled security(14.6%) and errors in trust list management (64.6%).

IV. Guidelines

To help secure ICSs, several guidelines have been published. The *National Institute of Standards and Technologies* (NIST) issued the comprehensive *Guide to Industrial Control Systems Security* [24] that provides a global picture of ICS security both in technical and organizational terms. A similar guideline is available from the German BSI [25].

In analogy to the IT world, the US *Computer Emergency Response Team* (CERT) publishes alerts and advisories concerning ICSs [26]. In line with this practice, manufacturers of ICSs have started to publish advisories and security bulletins [27][28][29].

From a technical perspective, secure coding practices for ICSs are emerging and collected in an open-source effort [30].

V. Conclusion and Future Work

The notion of PLCs being insecure by design is a recurrent theme in all presented work, the weakest links being a lack of authentication mechanisms and insecure protocols. OPC UA, when properly implemented and set up is the exception to the rule. Mechanisms like Intrusion Detection Systems [31] can help harden industrial systems. However, this is only reasonable after basic security mechanisms like authentication and secure protocols are put in place. Forensics [32][33] provides the cornerstone for not experiencing the same attack several times and helps to build vulnerability databases. Secure coding practices promote a *defense-in-depth* approach and help to reduce attack surfaces once the adversary gained access. However, they are only an additional layer of protection and cannot compensate for the aforementioned weaknesses.

As of today, many of the PLCs in the field are not or are insufficiently protected. Future work will thus be twofold: Targeting existing devices that have many of the vulnerabilities presented here and finding means to mitigate these in newer devices. The latter will mainly need to find fast and at the same time efficient cryptographic algorithms. While this is achievable with hardware-based symmetric encryption as [34] show, effective software-based solutions are still to be researched. For the first, however, our recommendation is to find and standardize ways of penetration testing in ICSs, such as Metasploit [35] from the IT world. This can then also be used to automatically check assets against newly discovered vulnerabilities. However, due to the variety of vendor-specific protocols and operating systems, this will be a demanding task.

References

[1] R. Langner, "Stuxnet: Dissecting a Cyberwarfare Weapon," *IEEE Security Privacy*, vol. 9, no. 3, pp. 49–51, 2011.

[2] R. M. Lee, M. J. Assante, and T. Conway, "Analysis of the Cyber Attack on the Ukrainian Power Grid," SANS Industrial Control Systems, Tech. Rep., 2016.

[3] K. Hemsley and R. Fisher, "A History of Cyber Incidents and Threats Involving Industrial Control Systems," in *Critical Infrastructure Protection XII*. Cham: Springer International Publishing, 2018, pp. 215–242.

[4] O. El Idrissi, A. Mezrioui, and A. Belmekki, "Inadequacy of IT Approaches to Manage Cyber Security in ICS Context," in *Proceedings of the 12th International Conference on Soft Computing and Pattern Recognition (SoCPaR 2020)*. Cham: Springer International Publishing, 2021, pp. 678–689.

[5] T. Fleury, H. Khurana, and V. Welch, "Towards a taxonomy of attacks against energy control systems," in *International Conference on Critical Infrastructure Protection*. Springer, 2008, pp. 71–85.

[6] A. Flowers, S. Smith, and A. Oltramari, *Security Taxonomies of Industrial Control Systems*, 08 2016, pp. 111–132.

[7] Z. Drias, A. Serhrouchni, and O. Vogel, "Taxonomy of attacks on industrial control protocols," in *2015 International Conference on Protocol Engineering (ICPE) and International Conference on New Technologies of Distributed Systems (NTDS)*, 2015, pp. 1–6.

[8] D. Beresford, "Exploiting Siemens Simatic S7 PLCs," *Black Hat USA*, vol. 16, no. 2, pp. 723–733, 2011.

[9] H. Wardak, S. Zhioua, and A. Almulhem, "PLC access control: a security analysis," in *2016 World Congress on Industrial Control Systems Security (WCICSS)*, Dec 2016, pp. 1–6.

[10] A. Ayub, H. Yoo, and I. Ahmed, "Empirical Study of PLC Authentication Protocols in Industrial Control Systems," in *Fifteenth IEEE Workshop on Offensive Technologies (WOOT)*, 2021.

[11] R. Spenneberg, M. Brüggemann, and H. Schwartke, "Plc-blaster: A worm living solely in the plc," *Black Hat Asia*, vol. 16, pp. 1–16, 2016.

[12] E. Biham, S. Bitan, A. Carmel, A. Dankner, U. Malin, and A. Wool, "Rogue7: Rogue engineering-station attacks on S7 Simatic PLCs," *Black Hat USA*, 2019.

[13] K. Wang, J. J. Parekh, and S. J. Stolfo, "Anagram: A content anomaly detector resistant to mimicry attack," in *International workshop on recent advances in intrusion detection*. Springer, 2006, pp. 226–248.

[14] H. Yoo and I. Ahmed, "Control logic injection attacks on industrial control systems," in *IFIP International Conference on ICT Systems Security and Privacy Protection*. Springer, 2019, pp. 33–48.

[15] S. Senthivel, S. Dhungana, H. Yoo, I. Ahmed, and V. Roussev, "Denial of engineering operations attacks in industrial control systems," in *Proceedings of the Eighth ACM Conference on Data and Application Security and Privacy*, 2018, pp. 319–329.

[16] B. Lim, D. Chen, Y. An, Z. Kalbarczyk, and R. Iyer, "Attack Induced Common-Mode Failures on PLC-Based Safety System in a Nuclear Power Plant: Practical Experience Report," in *2017 IEEE 22nd Pacific Rim International Symposium on Dependable Computing (PRDC)*, Jan 2017, pp. 205–210.

[17] S. Kalle, N. Ameen, H. Yoo, and I. Ahmed, "CLIK on PLCs! Attacking Control Logic with Decompilation and Virtual PLC," *Proceedings 2019 Workshop on Binary Analysis Research*, 2019.

[18] G. Bonney, H. Höfken, B. Paffen, and M. Schuba, "ICS/SCADA security analysis of a Beckhoff CX5020 PLC," in *2015 International Conference on Information*

Systems Security and Privacy (ICISSP), Feb 2015, pp. 1–6.

[19] T. H. Morris and W. Gao, "Industrial control system cyber attacks," in *1st International Symposium for ICS & SCADA Cyber Security Research 2013 (ICS-CSR 2013) 1*, 2013, pp. 22–29.

[20] OPC Foundation, "Practical Security Recommendations for building OPC UA Applications," 2018.

[21] Federal Office for Information Security, "OPC UA Security Analysis," 2016.

[22] F. Kohnhäuser, D. Meier, F. Patzer, and S. Finster, "On the Security of IIoT Deployments: An Investigation of Secure Provisioning Solutions for OPC UA," *IEEE Access*, vol. 9, pp. 99 299–99 311, 2021.

[23] A. Erba, A. Müller, and N. O. Tippenhauer, "Security Analysis of Vendor Implementations of the OPC UA Protocol for Industrial Control Systems," Apr. 2021.

[24] K. A. Stouffer, J. A. Falco, and K. A. Scarfone, "Guide to Industrial Control Systems (ICS) Security," 2011.

[25] BSI, "ICS Security Compendium," 2013, accessed: 2021-12-07. [Online]. Available: https://www.bsi.bund.de/SharedDocs/Downloads/EN/BSI/ICS/ICS-Security_compendium.html

[26] CISA, "Industrial Control Systems," accessed: 2021-12-07. [Online]. Available: https://us-cert.cisa.gov/ics

[27] Siemens, "Siemens Security News," 2023, accessed: 2023-02-01. [Online]. Available: https://new.siemens.com/global/en/products/services/cert/news.html#/posts

[28] Beckhoff, "IPC - Security Guideline - Advisories," 2023, accessed: 2023-02-01. [Online]. Available: https://infosys.beckhoff.com/english.php?content=../content/1033/ipc_security/976057355.html&id=

[29] Schneider, "Cybersecurity support portal," 2023, accessed: 2023-02-01. [Online]. Available: https://www.se.com/ww/en/work/support/cybersecurity/security-notifications.jsp

[30] PLC Security, "Top 20 PLC Secure Coding Practices," accessed: 2021-12-07. [Online]. Available: https://www.plc-security.com

[31] D. A. Yeager, L. Vega, A. Brown, H. Burke, J. Petersen, E. Rodas, J. Welsh, A. Wen, and M. Zaron, "Case Study: Architecting a Solution to Detect Industrial Control System Attacks," in *2020 IEEE Systems Security Symposium (SSS)*, 2020, pp. 1–8.

[32] I. Ahmed, S. Obermeier, S. Sudhakaran, and V. Roussev, "Programmable Logic Controller Forensics," *IEEE Security Privacy*, vol. 15, no. 6, pp. 18–24, November 2017.

[33] M. Cook, I. Stavrou, S. Dimmock, and C. Johnson, "Introducing a forensics data type taxonomy of acquirable artefacts from programmable logic controllers," in *2020 International Conference on Cyber Security and Protection of Digital Services (Cyber Security)*, June 2020, pp. 1–8.

[34] M. Skuballa, A. Walz, H. Bühler, and A. Sikora, "Cryptographic Protection of Cyclic Real-Time Communication in Ethernet-Based Fieldbuses: How Much Hardware is Required?" in *26th IEEE International Conference on Emerging Technologies and Factory Automation (ETFA)*, 2021, pp. 1–7.

[35] Rapid7, "Metasploit," 2023, accessed: 2023-02-01. [Online]. Available: https://www.metasploit.com

PIV of Flow Convection Induced by Solar Sphere to Generate Power

Hassan Abdulmouti
Associate Professor, Department of Mechanical Engineering Division
Sharjah Men's College, Higher Colleges of Technology
Sharjah, UAE
e-mail: habdulmouti@hct.ac.ae, alternative e-mail: hassanabujihad@hotmail.com

Abstract—**Flow visualization including image processing and Particle Image Velocimetry (PIV) measurements is used to evaluate the flow velocity field of the acrylic solar sphere filled with oil. This generated convection flow is affected by the thickness of the sphere, the sphere size as well as the temperature. The function of the acrylic sphere which is a modern technology of concentrated photovoltaic is to gather the energy from the sun and then concentrate it in a compact area like a focal spot. This focal spot is placed and put directly above a multi-junction device that acts as a concentrator cell appliance. This appliance directly can generate an enormous rate of power that is utilized to generate more electricity rather than the power that normal photovoltaic panels (PV) can produce. The acrylic sphere is used also for a lot of industrial applications. This research paper aims to investigate the characteristics of the flow inside the sphere and explore the sphere thickness, the sphere size, and the sphere temperature effect on the flow velocity of the fluid motion. Moreover, the purpose of this study is to clarify the relationship between these parameters in order to achieve greater efficiency for power generation, therefore, improve its performance. The results showed that the sphere thickness, the sphere temperature, and the sphere size significantly change the flow structure value. It is found that the velocity increase as the sphere thickness decreases. Hence, the efficiency of the sphere increases when using lower acrylic sphere thickness, higher size, and lower liquid temperature. The output power and efficiency of the solar sphere increase with lower sphere thickness. The thinner the thickness of the acrylic layer, the higher the sunlight absorbed by the acrylic photons. Subsequently, the higher the output power, which results to get higher the efficiency.**

Keywords- PIV; Solar, Flow; Concentrate; Energy.

I. INTRODUCTION

Several threats to food security, water resources, and the environment have appeared and expanded, resulting in a strong demand for technological advancements in the energy sector. This is also reinforced by data demonstrating the exponential rise in power and energy research over the previous few decades. Parallel to this, energy demand is expected to rise to a high between 2016 and 2040 [1]-[3].

Despite the global expansion of trends favoring energy conservation, green energy supply, and low environmental effect, producing enough energy to meet global demand remains a tough task. The most difficult task remains to deliver power from clean sources, such as renewable energy, which can lead to a reduction in the use of fossil fuels [4]-[6].

Renewable energy's capacity to meet the world's energy needs is promising, and it has the potential to reduce reliance on traditional energy sources, such as oil and uranium. Because traditional energy sources are both expensive and harmful to the environment, renewable energy sources provide a competitive option for power generation. Among the different renewable energy sources, solar energy is the most abundant and easily available to cover human electrical usage. As a result, numerous researchers have been looking for novel ways to gather solar energy and use it as a renewable energy source [7]-[9].

The efficiency of the solar sphere which is a new technology is significantly higher than normal solar panel collector components. Following the introduction of solar panels, many sun lights concentrated systems have arisen. One of these is the solar sphere collector, which is more efficient than traditional solar panels. Other benefits of the solar sphere include a) the ability to convert solar energy directly into electrical energy, b) solar radiations can be collected at any angle, unlike conventional solar panels, which must be integrated with a tracking system in order to capture the maximum amount of solar radiations, and c) their size is smaller compared to conventional solar panels, which occupy larger areas for relatively the same amount of solar radiations [10][11].

The concentrated solar sphere system consists of an acrylic sphere filled with a fluid that absorbs sunlight and transforms it directly into electrical energy. This is accomplished by concentrating solar energy from all across the sphere and focusing it in a small focused area on a high-efficiency solar cell. A pyranometer is used to measure sun irradiance, while a multimeter is used to measure the electrical power generated by the solar cell.

The solar sphere collector technology was already investigated and tested in our past study. The trials included different spherical materials, sphere shapes, and sphere sizes, as well as different fluids. Numerous tests were conducted at various circumstances throughout the year in order to acquire and assess the power production and related efficiency corresponding to the various parameters compared to the traditional solar panel (PV). Based on prior studies, the solar sphere displayed superior power production and efficiency when compared to the standard solar panel for the various parameters that were tested. In addition to this finding, the following conclusions were reached:

1. When compared to other shapes using the same fluid, the entire spherical produced the most power. The power was nearly four times that of a regular solar panel.

2. Increasing the size of the solar sphere increases power production and the associated efficiency.

3. In terms of the fluid used to fill the solar sphere, oil had a maximum power output that was 1.5 times that of alcohol. Alcohol, on the other hand, produced double the power output of water, with air producing the least power production of the fluids, tested [12]-[15].

Convection flow is the flow of fluid within the solar sphere, and it is found in a variety of applications, such as oil and natural gas transport, column reactor, air-sea gas transfer, ship hydrodynamics, boiling heat transfer, and bubble column reactors [16]-[19]. Researchers are interested in such convection flows because they occur in many engineering applications and numerical research and experiments aid in visualizing and understanding the parameters involved in this movement. Understanding and clarifying the flow parameters assist in defining the flow behavior and investigating the efficacy of the flow to improve the flow behavior towards the region of interest of the application at hand [20]-[22].

The goal of this research is to examine and comprehend the properties of flow in the solar sphere through the analysis of the velocity of fluid flow within the solar sphere using PIV. Furthermore, to investigate the influence of sphere thickness, sphere size, and sphere temperature on fluid motion and velocity. Furthermore, in order to increase the performance of the system to generate power and then maximize its efficiency, the link between these factors must be well defined.

In Section 2 of this paper the experimental apparatus and methods were illustrated. While in Section 3 the flow visualization and image processing results, including the PIV system and data analysis were explained. The conclusion was added in Section 4.

II. Experimental Methods and Apparatus

Figure 1 depicts the setup of the intended experiment. The acrylic (Plexiglas) solar sphere (No.1) material gives optical access, to the flow inside the sphere, to facilitate observation as well as image processing via PIV measurements and permits the collection of the solar radiation necessary to be focused. The sphere is full of frying oil (sunflower oil). The sphere is supported by a stainless-steel stand (No. 2).

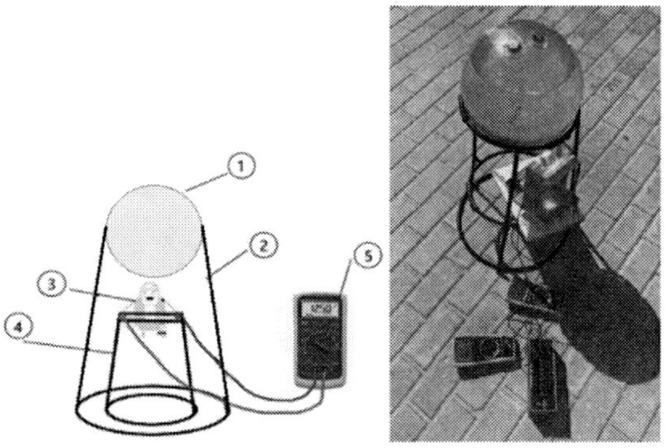

Figure 1. The setup for the solar sphere experiment.

A multi-junction concentrator solar cell (No. 3) is mounted on a stand (No. 4) beneath the solar sphere and is directly linked to a multi-meter (No. 5) to measure power production by measuring voltage and current ampere.

The solar sphere collects solar energy from the sun and focuses it on a single point. This focus point is located on top of a collector device, which is a multi-junction cell. The magnifying lenses in the multi-junction cell arrangement focus the solar radiations onto a focused region in the cell. This concentrated region is linked to a heat sink. The multi-junction cell is further subdivided into sub-cells, each of which is in charge of converting distinct components of the light into electricity. Because this device can withstand high temperatures and can help in radiation resistance, it was chosen to be utilized in this experiment to capture the concentrated solar energy of the focus point. The multi-junction device is connected to a multi-meter, which is used to read the output current and voltage. The experiment setup utilized to investigate flow characterization using flow visualization and image processing via PIV measurements consists of the apparatus shown in Figure 2 which is listed below:

1- As a light source, a Mini Diode-pumped solid-state laser (DPSS) Model #SM-SEMI-2W, which is a double pulse laser (also known as PIV laser), is employed. It employs two pulse lasers to emit laser beams via an optical beam combiner and a light path exit. The wavelength of the laser is 532 nm.

2- The CCD scientific class digital camera, Model #SM-CCDB2M25, records the visualized flows. This camera uses a double exposure mode that is coordinated with the double laser pulses and has a 50 mm f/1.4, F-Mount Lens, Model #SM-LENS5014. An external trigger initiates the capture. The frame grabber then sends both collected pictures to computer memory in real-time. A synchronizer generates trigger signals that are properly synced with the double-pulsed laser. The camera has a resolution of 1620×1220 (2M) and can take 25 frames per second (or 12.5 image pairs). The minimum exposure time interval in PIV mode is 200ns.

3- The synchronizer with USB cable, Model #MicroPulse 825 (MicroPulse725), provides cycle pulse trigger signals via the internal time base while also creating numerous delayed trigger signals via internal time-delay channels. The synchronizer is used to regulate the laser, digital camera(s), and frame grabber, ensuring that all of these diverse pieces function in perfect synchronicity.

4- A computer that is used to store picture data that has been captured by the frame grabber. It is then used by the Particle Image Velocity measurement system software to compute, display, and save the velocity field in real time.

5- The PIV system, which is 2D PIV software with particle image capture and velocity analysis, employs a laser as an independent illumination source that may be employed with or without the synchronizer. If the synchronizer is not accessible, the laser's internal synchronization can be used throughout the optical path and laser energy setup. Also employed is a high-resolution 2D2C PIV with multi-pass, multi-grid, window deformation method, and boundary deformation parameters. The frame grabber occupies a regular PCI (PCI-E) slot on the computer. The acquisition board's interface is primarily a 26-pin CamLink standard digital camera connection interface. Digital cameras with a standard CamLink interface are used in the Microvec PIV system. It connects to the frame grabber through three 10-meter signal lines, retains trigger signals to synchronize the digital cameras with the pulse laser, and then links to the synchronizer's output interface via coaxial signal cables via TTL trigger, which interfaces with the camera. When the synchronizer is provided as part of the PIV system, the laser must be set to external synchronous mode (where the synchronizer's 4-way delay signals are output to controls of the corresponding two sets of the laser flashlamp and Q-switch) while the digital camera is set to PIV work mode.

6- Fluorescent spherical tracer particles Model #MV-H1020 with a diameter of 7 μm (mean 10 μm) and a density of 1.04 - 1.06 g/cc are used as tracer particles to see the flow and determine velocity vectors.

As previously stated, the PIV measurements are employed in this study to produce the findings, which collect and calculate the velocity/velocity vectors of the flow inside the solar sphere, including flow visualization and image processing. To perform the tests within the lab, a lighting arrangement of 500 W Halogen lights is employed as a simulator instead of the sun. The laser is utilized with a black back sheet background to improve visibility and take high-quality photographs of the experiments for flow visualization and velocity measurements.

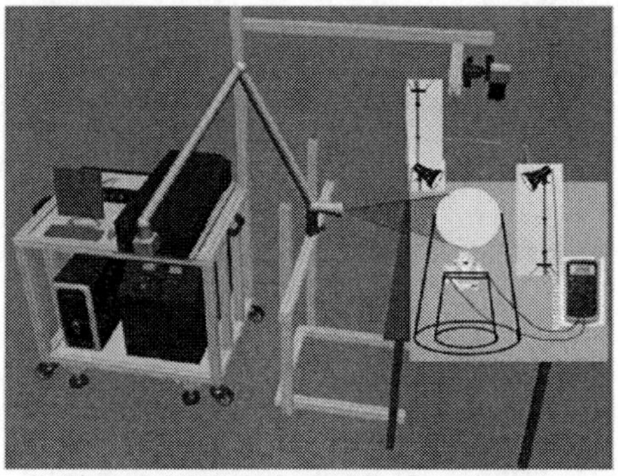

Figure 2. PIV Experimental Configuration

III. Flow Visualization and Image Processing Results, Including the PIV System and Data Analysis

Particle Image Velocimetry (PIV) is an optical flow imaging technique. It made use of the acquired instantaneous velocity measurements as well as the related fluid characteristics. Tracer particles must be introduced into the fluid to help in the capture of velocity vectors. The velocity vectors will be generated by lighting the region of interest with the laser and then monitoring the seeded particles. The tracer or seeded particles are supposed to follow the motion of the fluid, which is evenly dispersed throughout the flow field, and the interrogation window, which has a distinct speed. When the tracer particles are in motion, their flow is caught by sequential imaging, which is then processed for additional cross-correlation, allowing the observed flow's speed and direction (the velocity field) to be calculated. Furthermore, additional processing may be performed using the flow vortices, flow field parameter distribution, speed lines, and flow lines. The configuration of a typical PIV system typically consists of four basic physical components. The digital CCD or CMOS camera is the first. The laser and its optical arrangement, which restricts the physical lit region of interest, are the second component. The third component is the synchronizer, which functions as an external trigger for regulating and timing both laser and cameras. As previously stated, the seeding particles are the fourth component. Of course, all of these components must be employed in conjunction with the fluid under investigation. The laser can be linked to specific lenses to transform the laser beam into a sheet or line ray. Finally, the optical pictures collected will be processed using specialist PIV software. The tracer particles following the movement of the fluid are emitted by the pulsed laser within the known time interval t, and the sheet light illumination by the lens group records the particles' instantaneous position on the CCD chip. If we know the displacement change of the same particle micelle at two times t1 and t2 from the recorded particle picture, we can calculate the velocity of the particle group at t1 using the definition of velocity, as indicated in the formula below [23]-[29].

$$v = \lim_{\Delta t \to 0} \frac{\Delta s}{\Delta t}$$

To begin analyzing the obtained picture, the idea of interpretation area must be defined. It refers to a square picture of a specific size in a specific location in the image, and the speed is acquired by executing signal processing in the interpretation area. Assume the system captures images 1 and 2 in Figure 3 at two different times, t_0 and t_{0+t}. Obtaining two interpretation regions of the same size, f(m,n), in the same spot in the picture. where (m, n) denotes the relative location of f and g in images 1 and 2, respectively. Processing f and g yields the appropriate displacement S of the interpretation area, as indicated in Figure 3.

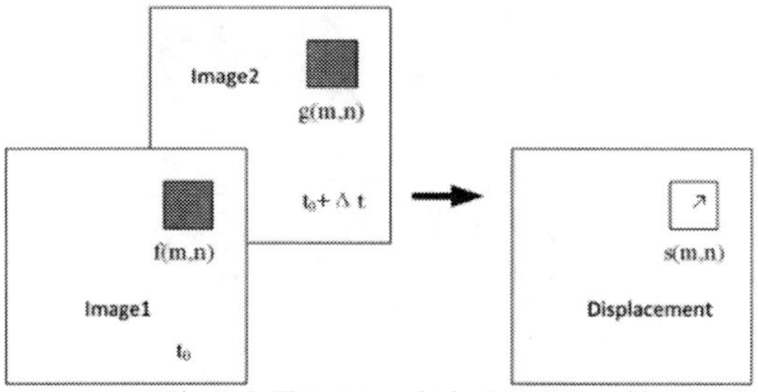

Figure 3. The system obtains images 1 and 2.

Figure 4 depicts the link between the digital signal transfer function and the interpretation regions f, g, and the displacement vector (the uppercase letters in the figure correspond to the lowercase Fourier transform). TSI's INSIGHT 4G® software is also utilized for picture and correlation analysis. The technique begins by analyzing the pictures using 6464 Interrogation Areas (IAs) using a Fast Fourier Transform (FFT) correlator, resulting in a 50% overlap of the interrogation areas. The correlation peak is then examined to acquire the findings of the final questioning region of 1616 pixels in size. The Gaussian curve-fitting approach with subpixel precision is used to determine this. The median technique, one of the post-processing methods available in the TSI analysis program, is used to remove spurious vectors from velocity fields. The rejection rate of the vectors is roughly 3%. The Gaussian-weighted technique replaces these rejected vectors with their neighbors [30]-[34].

In the applied image processing, an average duration of 10 μs consecutive frames (between 2 consecutive frames) is utilized to discover the values of the velocity of the flow inside the spheres of 10 cm 15 cm, and 30 cm diameters, and for different thicknesses of 3, 4, 5, 6 and 8 mm. These experiments are conducted for 3 different temperature rates of 35, 40, and 45°C. The velocity is determined and then computed for all these spheres/cases. Image processing calculates the flow velocity estimate at roughly 10,000 vectors of the flow pictures. The velocity experiments are carried out for the entire sphere. For each example, the temperature, voltage, and current ampere were measured, hence the output power was determined.

Figure 5 depicts a sample of images of the sphere with a diameter of 15 cm and a thickness of 8 mm, whereas Figure 6 depicts a sample of flow images for the experimental acrylic sphere with a diameter of 10 cm and a thickness of 4 mm illuminated by laser for flow visualization and further image processing for velocity size computations.

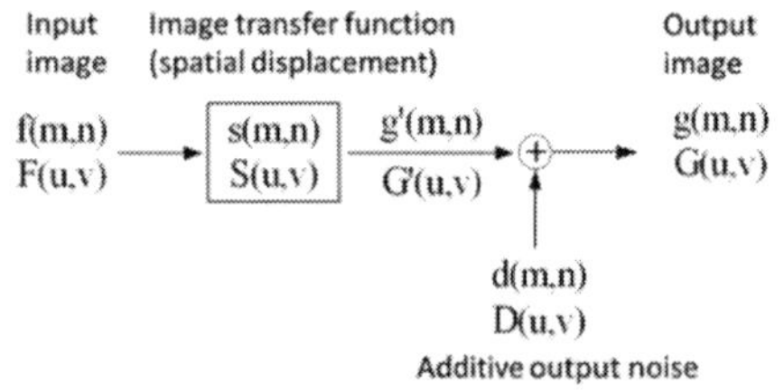

Figure 4 The connection between the interpretation regions f and g

Figure 5. Images of a sphere with a diameter of 15 cm and 8 mm thickness

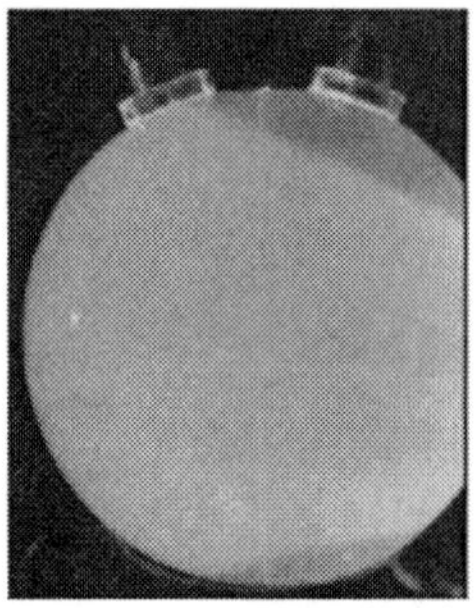

Figure 6. Images of an experimental laser-illuminated acrylic sphere with a diameter of 10 cm and a thickness of 4 mm

Figure 7 shows a PIV vector plot with an instantaneous velocity field (vectors sample) of the flow within the sphere of 15 cm diameter and 3 mm thickness at a temperature of 35°C measured by PIV between the first frame and frame number 200 in order to easily distinguish the flow inside the sphere since the flow is quite sluggish. Figure 8 illustrates a sample of the immediately dispersed velocity vector of the flow in a sphere of 15 cm diameter and 5 mm thickness measured by PIV between the first and 80th frames at a temperature of 35°C. Figure 9 illustrates a sample of the immediately dispersed velocity vector of the flow in a sphere of 15 cm diameter and 8 mm thickness measured by PIV between the first and frame number 200 at a temperature of 35°C. Figure 10 demonstrates the result of the flow visualization and vorticity of the flow in the sphere of 10 cm diameter and 6 mm thickness measured by PIV between the first and the 40th frame at a temperature of 45°C.

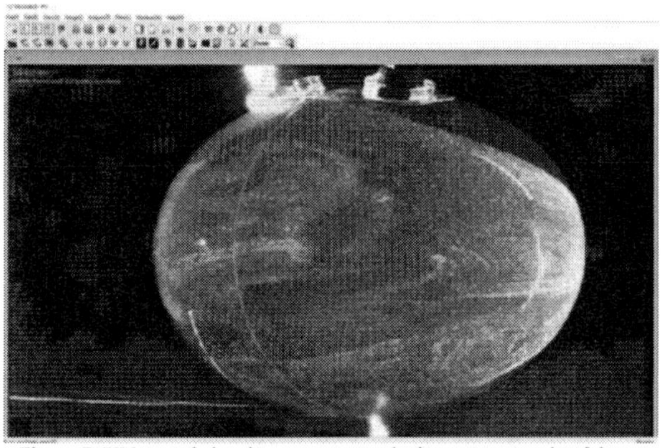

Figure 7. PIV recorded an instantaneous velocity vector sample of the flow in a sphere of 15 cm diameter and 3 mm thickness (1-200 frame) at a temperature of 35°C.

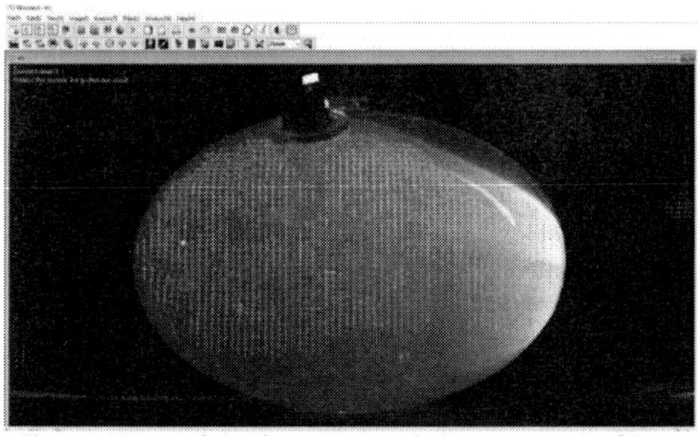

Figure 8. PIV recorded an instantaneous velocity vector sample of the flow in a sphere of 15 cm diameter and 5 mm thickness (1-80 frame) at a temperature of 35°C.

The velocity of the fluid (particles) at the top of the sphere is clearly higher than the velocity of the seeded particles in other regions of the fluid motion, as shown by these figures. Because there are many interaction types between the flow of the liquid and the upper part of the inside wall of the sphere, the flow generated on the top of the sphere is considered an appropriate technique that aids in the detailed investigation of the flow regime during the flow motion from the bottom of the sphere to the top of the sphere. Furthermore, the liquid flow characteristic at the top region does not behave similarly to other flow regions. This is because the fluid flow on the top reaches the upper wall and reacts with it by mirroring the direction downward. As a result, the upper half of the spherical wall guides, directs, and reflects the fluid flow, whereas the flow in other places delays the fluid. Actually, the interaction between the fluid flow and the spherical wall is extremely evident. As a result, this inquiry and its accompanying data assist in determining the entrainment flow in the downward direction. This also encourages research into the relative velocity expression and its reaction and flow behavior to the encirclement liquid.

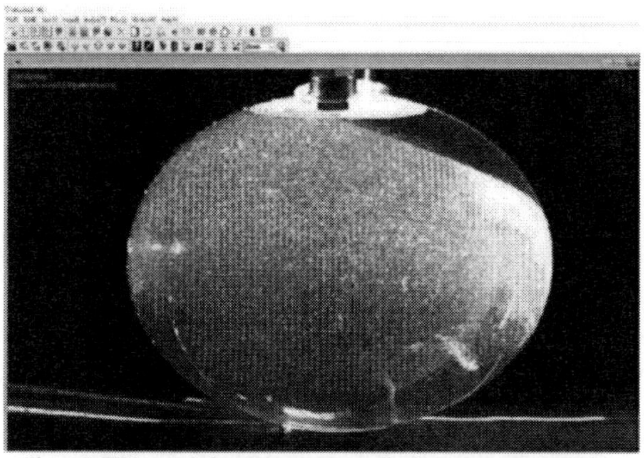

Figure 9. PIV recorded an instantaneous velocity vector sample of the flow in a sphere of 15 cm diameter and 8 mm thickness (1-200 frame) at a temperature of 35°C.

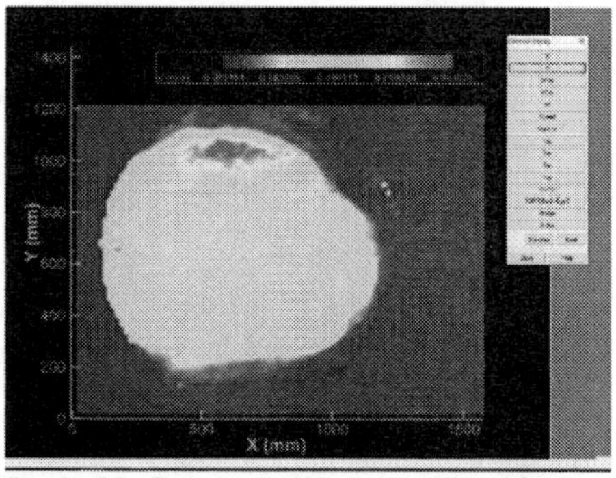

Figure 10 Result of the flow visualization and vorticity of the flow in the sphere of 10 cm diameter and 6 mm thickness (1-40 frame)

The complete explanation of the flow process within the sphere may be presented and summarized using the resulting figures as follows. The produced convection flow pattern is equally harmonic around the sphere center, where the movements were primarily caused inside the fluid as a result of the tendency of the fluid with higher temperature and lower density to rise and the fluid with lower temperature and higher density to sink with the effect of gravity, resulting in heat transfer. Furthermore, the flow is observed to be steady and uniform around the center, with the exception of the region of the top surface where the fluid flow impinges and reflects. The rising flow around the center is steered by the sphere's wall, resulting in a significant fluid flow formation towards the upper region of the sphere. The flow evolves as follows: the maximum velocity for the upward flow momentum is obtained towards the upper top region of the sphere. When it is reflected downhill on the wall, the orientation of the upward flow changes quickly from an upwards direction to a downward direction flow. Following the formation of two circular liquid currents around the center, and after a specific amount of time has passed, those circular currents encourage the center of the sphere with a scale circulation throughout the complete fluid layers.

Figures 11, 12, and 13 show the relationship between the sphere thickness of 10 and 14, and 30 cm diameter and the fluid velocity for different temperatures of 35, 40, and 45°C and for a sphere thickness of 3, 4, 5, 6, and 8 mm. The fluid velocity is calculated to be the average velocity of the top region under the wall surface. Because this is the more complicated area where the fluid has many interactions with the upper surface.

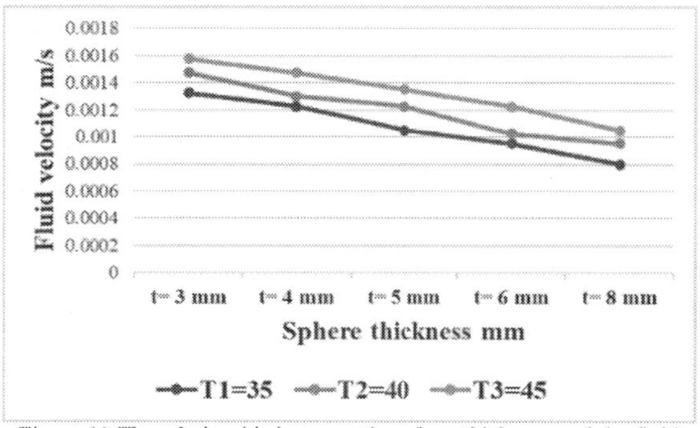

Figure 11 The relationship between the sphere thickness and the fluid velocity for different temperatures and for a sphere diameter of 10 cm

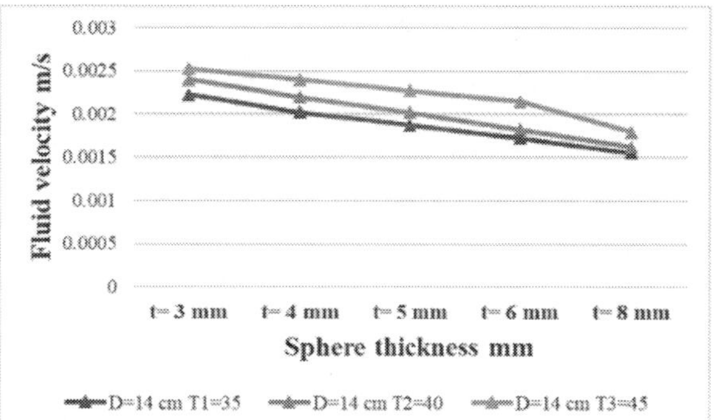

Figure 12 The relationship between the sphere thickness and the fluid velocity for different temperatures and for a sphere diameter of 14 cm

Figure 14 illustrates the relationship between the sphere thickness of 10 cm diameter with the output power and efficiency for a temperature of 35°C and for a sphere thickness of 3, 4, 5, 6, and 8 mm. The current-voltage characteristics were plotted for each sphere thickness of 3, 4, 5, 6, and 8 mm separately. The optimum operating point is at the maximum power point. The values of these maximum power points were accumulated in Figure 14, It is clear from this figure that the output power and efficiency increase when using lower acrylic thickness. Hence, using a 3 mm sphere thickness is the best.

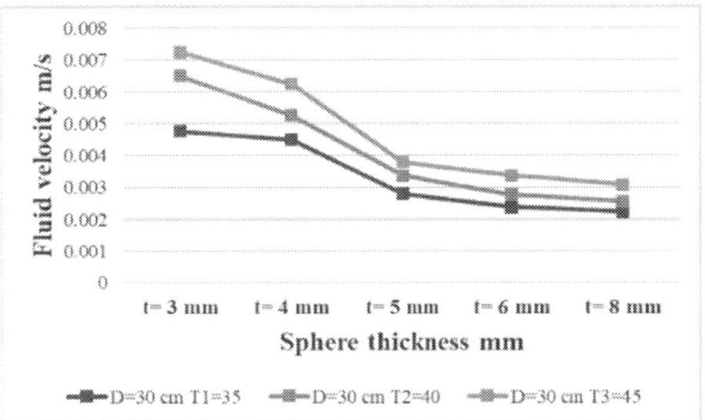

Figure 13 The relationship between the sphere thickness and the fluid velocity for different temperatures and for a sphere diameter of 30 cm

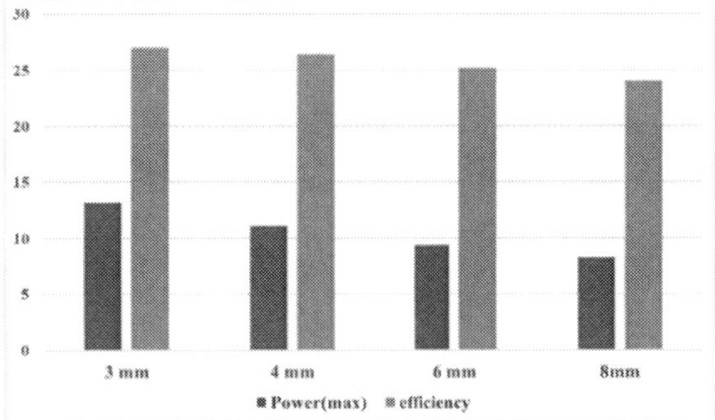

Figure 14 the relationship between the sphere thickness of 10 cm diameter and the output power and the efficiency for a temperature of 35°C

According to the figures, the fluid flow approaches the top spherical surface as the velocity increases. Furthermore, when comparing the velocity vectors, it is discovered that when the spherical thickness lowers, the velocity increases. Whereas the average velocity of the 4 mm thickness sphere is nearly double that of the 6 mm thickness sphere in the region of the top surface of the sphere, and the average velocity of the 6 mm thickness sphere is nearly double that of the 8 mm thickness sphere in the region of the top surface of the sphere. Also, as the temperature increases the fluid velocity increases. Moreover, when the size of the sphere increases the fluid velocity also increases. Furthermore, it is acknowledged that the output power and the efficiency of the 4 mm thick sphere are greater than that of the 6 mm thick sphere. In reality, as the thickness of this acrylic solar sphere rises, the amount of solar radiation received is influenced and limited, becoming less. As a result, the output power is decreased. Hence, the greater the thickness of the acrylic, the less electricity is produced by the sun. When the thickness is reduced, more power is produced, resulting in improved efficiency. As a result, the thinner the acrylic coating, the more solar light is absorbed by the PV photons. As a result, the larger the output power, the higher the efficiency.

Finally, the behavior of flow motion within the solar sphere was explored in an attempt to improve power generation performance. The examination in this research effort was carried out to understand the influence of sphere thickness on fluid flow, which is elucidated in this document, in order to increase fluid flow performance and, hence, maximize the related efficiency. The results revealed that the thickness of the sphere considerably alters the flow structure and fluid velocity value. As a result, a thinner sphere should be employed to increase the efficiency of the solar sphere. Furthermore, the thinner the acrylic coating, the more sunlight is absorbed by the PV photons. As a result, the larger the output power, the higher the efficiency.

IV. Conclusions

The flow visualization and image analysis using PIV measurements on the fluid flow within a solar sphere and the velocity motion of the filled oil fluid were carried out in this study to optimize the flow behavior. The flow velocity and its influence on sphere thickness are calculated. The collected findings demonstrated a clear influence of spherical thickness and temperature on fluid flow. The following is an overview of the key conclusions:

1. The output power and efficiency increase when using lower acrylic thickness. Hence, using a 3 mm sphere thickness is the best.
2. When the temperature increases the fluid velocity increases and when the size of the sphere increases the fluid velocity also increases
3. The fluid flow velocity was discovered to grow and climb in value to the upper surface wall of the sphere, the mean velocity magnitude was determined, and it was discovered to be greater in the region of the top surface than the starting velocity in the other regions.
4. The fluid velocity data showed that the velocity increases as the thickness of the sphere decreases, and the mean velocity magnitude progressively increases as one rises within the sphere.
5. The output power and the associated efficiency of the solar sphere increase with respect to the properties of lower sphere thickness. The thinner the thickness of the acrylic layer, the higher the sunlight absorbed by the PV photons. Subsequently, the higher the output power, which results to get higher the efficiency.

Acknowledgment

The study described in this publication was supported by the Internal Interdisciplinary Research Grant No. (212194), Fund number 113180. HCT stands for High Colleges of Technology in the United Arab Emirates.

References

[1] H. Abdulmouti, K. Ali, A. Ali, M. Ali, S. Abdullah, R. Abdalla, "Smart Innovation Applications for a GreenHouse Using Sustainable and Renewable Energy in the UAE," Publisher: IEEE. DOI: 10.1109/ICASET.2018.8376782, Electronic ISBN: 978-1-5386-2399-2, Print on Demand (PoD) ISBN: 978-1-5386-2400-5, IEEE Xplore Digital Library: 11 June 2018.

[2] H. Abdulmouti, Z. Skaf, and S. Alblooshi, "Smart Green Campus: The Campus of Tomorrow," 2022 Advances in Science and Engineering Technology International Conferences (ASET), 2022, pp. 1-8, DOI: 10.1109/ASET53988.2022.9735087, Publisher: IEEE. Date Added to IEEE Xplore: 18 March 2022.

[3] O. Edenhofer, R. Pichs-Madruga, Y. Sokona, K. Seyboth, S. Kadner, T. Zwickel, & P. Matschoss, (Eds.), "Renewable energy sources and climate change mitigation: special report of the intergovernmental panel on climate change," Cambridge University Press, [Accessed 15 March. 2017].

[4] H. Abdulmouti, "Passive Cooling Module to Improve the Solar Photovoltaic (PV) Performance," The 4th Int. Conf. on Environment, Chemical Engineering & Materials. Athens, Greece, December 27-29, 2022. Published on WSEAS Transactions on Power Systems, vol. 18, pp. 11-17, 2023.

[5] Green Facts, "Forests & Energy: 2, What are the trends and prospects of energy supply and demand?," [Accessed 3 March. 2017].

[6] H. Abdulmouti, "Innovative Environment-Friendly Systems for a Modern Town," 12th International Conference on Sustainable Energy & Environmental Protection (SEEP'19), 18-21 Nov. 2019, UOS, Sharjah, UAE.

[7] H. M. Cooper, "Organizing knowledge syntheses: a taxonomy of literature review," Knowledge Society, 1, 104-126, [Accessed 3 March. 2017].

M. Marsal-Llacuna, J. Colomer-Llinàs, J. Meléndez-Frigola, "Lessons in urban monitoring taken from sustainable and livable cities to better address the Smart Cities initiative," Technological Forecasting and Social Change,Volume 90, Part B,2015,Pages 611-622, January 2015,

[8] V. Quaschning, "Renewable Energy and Climate Change," Berlin University of Applied Systems HTW, Germany, (Translator) Hedy Jourdan, A John Wiley & Sons, Ltd., Publication, 2010.

[9] Green Rhino Energy, "Concentrating Photovoltaics Solar Power," Retrieved May 06, 2016. http://greenrhinoenergy.com/solar/technologies/pv_concentration.php

[10] P. Salmerón Revuelta, S. Pérez Litrán, J. Prieto Thomas, "Design, Simulation and Implementation for Improving Power Quality," Distributed Generation, Academic Press, pages 285-322, ISBN 9780128032169, https://doi.org/10.1016/B978-0-12-803216-9.00008-0, 2016.

[11] H. Abdulmouti, "An Experimental Innovative Solar Sphere Design to Generate Electricity," 5th International Conference on Renewable Energy and Development (ICRED 2019), 20- 23 September 2019, Okinawa, Japan.

[12] H. Abdulmouti, A. Ahmed Aljasmi, M. Ali Almarzooqi, H. Alyasi, M. Jassim Khair, Y. Mohamed Almulla, A. Ahmed Almulla, "Generating Power from Innovative Solar Sphere," DOI: 10.1109/ICASET.2019.8714441, ISBN Information: Electronic ISBN: 978-1-5386-8271-5, Print on Demand (PoD) ISBN: 978-1-5386-8272-2, Date of Publisher: 16 May 2019, Publisher: IEEE Xplore.

[13] H. Abdulmouti, Producing Electricity by Concentrated Solar Energy," The 2nd International Conference on Advances in Energy Research and Applications (ICAERA'21), November 24 - 26, 2021, Seoul, South Korea.

[14] H. Abdulmouti, Z. Skaf, S. Alblooshi, S. Almheiri, "Exploring the Flow of Solar Sphere Using PIV," 5th Advances in Science and Engineering Technology multi-conferences (ASET 2023), International Conference on Sustainable Environment and Urban Infrastructure, "in press", 20-23 Feb. 2023.

[15] H. Abdulmouti, "Numerical Simulation of Bubble Convection in Two-phase Stratified Liquids," Multiphase Science and Technology by Begell House, Inc "USA", Volume 31, Issue 2, pp.:133–149 (2019), The Home for Science and Engineering, DOI: 10.1615/MultScienTechn.2019029995.

[16] H. Abdulmouti, "Experimental Measurement for Surface Flow Characteristics Generated by a Bubble Plume," The Journal of Flow Visualization and Image Processing, 22 (1–3), 39–58, 2015, Begell House "USA", The Home for Science and Engineering.

[17] H. Abdulmouti, "Experimental Measurements of Bubble Convection Models in Two-phase Stratified Liquids," Experimental Thermal and Fluid Science, Volume 83C, May 2017, pp. 69-77, (ELSEVIER), https://doi.org/10.1016/j.expthermflusci. 2016.12.010.

[18] H. Abdulmouti, "Bubbly Two-Phase Flow: Part II-Characteristics and Parameters," DOI: 10.5923/j.ajfd.20140404.01 American Journal of Fluid Dynamics, Volume 4, Number 4, 4(4): 115-180, December 2014, (Scientific and Academic Publishing).

[19] H., Abdulmouti, "Bubbly Two-Phase Flow: Part I-Characteristics, Structures, Behaviors and Flow Patterns," American Journal of Fluid Dynamics, Scientific and Academic Publishing, vol. 4, no. 4, 4(4), pp. 194-240. DOI: 10.5923/j.ajfd.20140404.03. December, 2014a.

[20] H., Abdulmouti, "The Measurements of Bubble Plume Structure Parameter," International Journal of Fluid Mechanics Research, Begell House "USA", The Home for Science and Engineering, 44(3): 1–19, 1064–2277/17/$35.00, 2017.

[21] H. Abdulmouti, "Measurements of Thermal Effect on Bubble Parameter," The 3rd Thermal and Fluids Engineering Conference (TFEC), March 4-7, 2018, Fort Lauderdale, FL, USA.

[22] H. Abdulmouti, "Particle Imaging Velocimetry (PIV) Technique: Principles, the typically used methods, classification and applications," Scholar's Press. ISBN-13: 978-3-639-51249-6, 6 March 2013.

[23] H., Abdulmouti, "2D- numerical simulation of surface flow velocity and internal flow structure generated by bubbles," Multiphase Science and Technology, Begell House (USA), the Home for Science and Engineering, 28 (2): 153–171, 0276–1459/16/$35.00 © 2016, DOI: 10.1615/MultScienTechn.2017018926, 2016.

[24] H. Abdulmouti, "Measurement of Flow Structures Induced by a Bubbly Plume Using Visualization, PIV, and Image Measurement," Scholar's Press, Saarbrücken, Germany, ISBN-13: 978-3-639-51490-2, 7, June 2013.

[25] R. Bastiaans, "Cross-correlation PIV : theory, implementation and accuracy," Compustion Technology, 4, 2000.

[26] H. Abdulmouti and M. Shinneeb, "Investigation of Free-Surface Flow Induced by a Bubbly Plume Using PIV," 2020 (ASET), Electronic ISBN: 978-1-7281-4640-9. ISBN: 978-1-7281-4641-6. Print on Demand (PoD), pp. 1-5, DOI: 10.1109/ASET48392.2020.9118203, Publisher: IEEE. Added to IEEE Xplore: 16 June 2020.

[27] H. Abdulmouti and E. Jassim, "Visualization and Measurements of Bubbly Two-Phase Flow Structure Using Particle Imaging Velocimetry (PIV)," European Scientific Journal (ESJ), /special/ edition No.3, ISSN: 1857 – 7431, pp. 197-208, June 2013.

[28] H. Abdulmouti. "Particle Imaging Velocimetry (PIV) Technique: Principles and Application," Yanbu Journal of Engineering and Science, 2.1, pp. 50-67, 2021.

[29]

[30] H. Abdulmouti, Y. Murai, J. Ohta, and F. Yamamoto, "Measurement of Bubble Plume Generated Surface Flow Using PIV," Journal of the Visualization Society of Japan, Vol. 21, No. 2, pp. 31-37, 2001.

[31] H. Abdulmouti, Y. Murai, J. Ohta, and F. Yamamoto, "PIV Measurement of Bubbly Flow Interaction with Water

Surface," Journal of the Visualization Society of Japan, Vol. 19, No. 2, pp. 209-210, 1999.

[32] H. Abdulmouti, Y. Murai, J. Ohta, and F. Yamamoto, "PIV Measurement of Surface Flow Induced by Bubble Curtain," Journal of the Visualization Society of Japan, Vol. 19. No. 1, pp. 239~242, 1999.

[33] H. Abdulmouti, Y. Murai, J. Ohta, and F. Yamamoto, "PIV Measurement and Numerical Analysis of Flow around Bubble Curtain," Japanese Society of Mechanical Engineer (JSME), No. 987-1, pp. 83-84, 1998.

[34] H. Abdulmouti, and T. Mohamed Mansour. "The Technique of PIV and Its Applications," 10th International Congress on Liquid Atomization and Spray Systems (ICLASS-2006), 27 Aug.-1 Sept. Kyoto, Japan.

An Architecture for Reliable Learning Agents in Power Grids

Eric MSP Veith
Carl von Ossietzky University Oldenburg
Research Group Adversarial Resilience Learning
Oldenburg, Germany
Email: eric.msp.veith@uol.de

***Abstract*—Agent systems have become almost ubiquitous in smart grid research. Research can be roughly divided into carefully designed (multi-) agent systems that can perform known tasks with guarantees, and learning agents based on technologies, such as Deep Reinforcment Learning (DRL), that promise real resilience by learning to counter the unknown unknowns. However, the latter cannot give guarantees regarding their behavior, while the former are limited to the set of problems known at design time. This paper presents a hybrid architecture that enables a learning agent to give guarantees about its behavior, making it suitable for usage in Critical National Infrastructures (CNIs), such as the power grid.**

***Keywords*—agent systems; reinforcement learning; trustworthy AI; resilience; power grid**

I. INTRODUCTION

Over the last years, agent systems and especially Multi-Agent Systems (MASs) [1]–[4] have emerged as one of the important tools to facilitate management of complex energy systems. As swarm logic, they can handle numerous tasks, such as maintaining real power equilibria, voltage control, or automated energy trading [5]. The fact that MASs implement proactive and reactive distributed heuristics allows to analyze their behavior and give certain guarantees, a property that has helped in their deployment.

However, modern energy systems have also become valuable targets. Cyber-attacks have become more common [6], [7], and establishing local energy markets, although being an attractive concept of self-organization, can also be to manipulation, e. g., through artificially created congestion [8]. Attacks on power grids are no longer carefully planned and executed, but also learned by agents, such as market manipulation or voltage band violations [9]. Thus, carefully designing software systems that provide protection against a widening field of adversarial scenarios have become a challenge, especially considering that complex, inter-connected Cyber-Physical Systems (CPSs) are inherently exploitable due to their complexity itself [10].

Learning agents, particularly those based on DRL, have gained traction as a potential solution: If a system faces unknown unknowns, a learning agent can devise strategies against it. In the past, researchers have published using DRL-based agents for numerous tasks related to power grid operations—e. g., voltage control [11]—, but the approach to use DRL for general resilient operation is relatively new [12], [13]. DRL—the notion of an agent with sensors and actuators that learns by "trial and error"—is at the core of many noteworthy successes, such as MuZero [14], with modern algorithms such as Twin-Delayed DDPG (TD3) [15], Proximal Policy Gradient (PPO) [16], and Soft Actor Critic (SAC) [17] having proved to be able to tackle complex tasks.

All modern DRL use deep Artificial Neural Networks (ANNs) at least for the policy (or multiple, e. g., for the critic). Actual parameter optimization is commonly done with gradient descent algorithms. However, these ANNs' architectures still need to be provided by the user, in addition the hyperparameters of the algorithm. No DRL agent is, therefore, a "deploy and forget" approach; careful tuning is usually required for a specific use case. Evolving these networks, or using genetic or evolutionary algorithms as an alternative entirely, has gained interest among scientist during the last years [18], [19].

However, these model-free algorithms themselves cannot give guarantees with regard to their behavior, which is important for deployment with high autonomy in any CNI. *Safe DRL* algorithms target this research gap, but currently, they learn inefficiently or explore insufficiently [20]. Moreover, Safe DRL does not (yet) tackle changes in tasks or environment, the problem of *Online Learning* [21].

In a very similar vein, learning agents for CNIs not only need to give guarantees, but they must also offer introspection. I. e., their strategies must be inspectable by humans. In the context of CNIs, this allows audition or certification, or testing and validation even at runtime. Finally, it is an important factor for acceptance. This introspection is provided by techniques of eXplainable Reinforcement Learning (XRL) [22]. However, the most common techniques, such as saliency maps, give only indirect interpretation and are useful for experts in the DRL domain, but not for practitioners in CNIs. Recent approaches to convert a DRL agent's policy network into a rule-based representation, e. g., as decision tree [23], will satisfy the outlined requirements, but are not part of an DRL agent architecture yet.

Therefore, we identify the following research gap: Learning agents are necessary for modern, complex CNIs, such as the power grid. In order to cope with the complexity of the power grid, changing actor behavior and, thus, changing marginal distributions, online learning must be explicitly considered. In addition, such a learning agent must be inspectable and provide guarantees. To this end, we propose a hybrid architecture in the Adversarial Resilience Learning (ARL) agent. It combines a learning agent with a rule-based agent. The learning agent's policy is constantly converted into rule sets, represented in the Boolean domain, in order to enable the benefits of XRL.

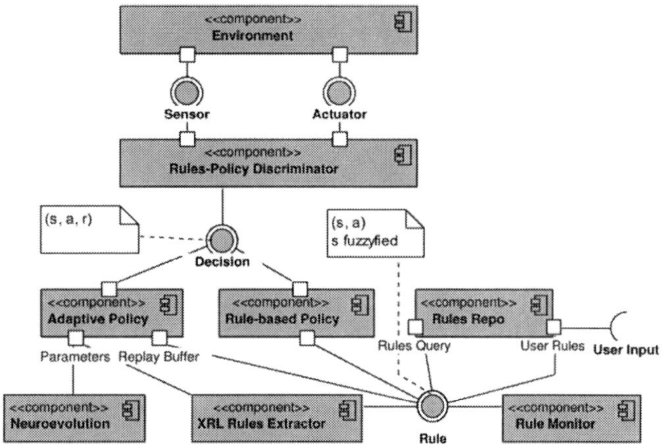

Figure 1. ARL Agent Architecture

However, the agent is also able to absorb obsolete rules when a better strategy is (reliably) devised, in order to maintain the agent's adaptability. Through this cycle of rule extraction (XRL) and rule consumption (online learning), the agent stays inspectable and validatable, providing the necessary security for deployment in CNIs, while still being able to learn and counter the unknown unknowns.

The remainder of this work-in-progress paper is structured as follows: Section II outlines the planned architecture and gives the rationale for each module, as well as their interfaces. In Section III, we outline our testing scenarios, which serve as overall validation of the architecture. Finally, we outline the next steps in Section IV.

II. Agent Architecture and Modules

The key design goal of the ARL agent architecture is to fuse a learning component based on DRL with the analyzability of a rules-based architecture. Therefore, an *Adaptive Policy* based on DRL, as well as a *Rule-based Policy*, are employed alongside each other. The DRL-based architecture will resort to off-policy algorithms, such as TD3 [15] and SAC [17]. This will simplify the extended usage of replay buffers, e. g., for rehearsal or offline learning.

This design iteration of the architecture does not consider MASs, therefore, the design follows more the premises of DRL-based agents and eschews the usual message inbox, journal, etc. that would normally be present for agents of a MAS. The resulting component design is depicted in Figure 1.

The agent receives data from the environment using *Senors*, which constitutes the agent's world state at time t: s_t. A sensor is a simple software interface that transfers sensor readings according to a simple mathematical space definition that allows for Boolean values, intervals of discrete numbers, n-dimensional real-valued spaces, and more complex types created as combinations thereof. We loosely follow the example of OpenAI Gym [24] and use the definition that we introduced in prior works [25].

Sensor readings are fed into the *Rules-Policy Discriminator*. This module is responsible for choosing the agent's actions, which stem either from the adaptive policy or the rule-based one. The discriminator gives precedence to behavior emitted from the rules-based policy, thus ensuring well-defined behavior whenever possible. If the rule-based policy cannot emit an action given the current sensor readings, the discriminator uses the action proposed by the adaptive policy. However, even if a matching rule can be emitted from the rule-based policy, the adaptive policy is still queried for an action. This way, $(s_t, a_t, \hat{r}_t)$ triplets can be fed to the replay buffer. The adaptive policy's action is preferred if it, checked against a world model, provides a higher reward. The triplet of state s_t—i. e., the current sensor readings—, the planned action a_t of the agent, as well as the reward $\hat{r}_t$ the agent expects, is an agent's *Decision*. The discriminator also propagates the actual reward from the last action, r_{t-1}, to the policies in order to update their replay buffer or rules assessment (adaptive and rule-based policy, respectively).

The rule-based policy uses a *Rules Repository* for storage. The repository uses Ternary Vector Lists (TVLs) [26] for efficient storage. TVLs represent systems of boolean equations as lists of disjoint vectors, where each vector represents the assignment of variables. In contrast to binary vectors, ternary vectors provide efficient storage by introducing a third symbol, such that a ternary value is defined as $tv \in \{0, 1, -\}$. The dash expands to both 0 and 1, so that the ternary vector $[1, -]^\top$ expands to two binary vectors, $[1, 0]^\top$ and $[1, 1]^\top$. Set operations on TVLs are well-defined and expand to the appropriate operations on Boolean values. The ARL agent will expand the notion of TVLs fuzzy logic in order to allow rules-based inference of actions based on sensor readings when those readings will seldom provide the exact values in $\mathbb{R}^n$ that are given in the (Boolean) rule sets.

A *Rules Monitor* supervises the repository: The agent still needs to be adaptive and learn, i. e., develop new strategies to situations unknown at design time. Therefore, the monitor will feed rules of known sensor readings back to the DRL policy's replay buffer. Should the adaptive policy have devised better rules, then this is merely the same problem off-policy DRL algorithms solve with their replay buffer, which contains triplets of $(s, a_1, r_1), (s, a_2, r_2) : r_1 < r_2$ and is a simple optimization problem. In the case of catastrophic forgetting [21] during online learning, the rules repository serves as a rehearsal device [21].

The combination of rules repository and rules monitor serves the training aspect of the agent: The adaptive policy will be trained using *Neuroevolution*, i. e., the ANNs are evolved during training, their architecture not provided beforehand. Usually, employing neuroevolutionary strategies reduces the sample efficiency considerably. However, the rules repository-monitor modules are able to serve as extended replay buffer, allowing for extended iteration over samples to use them during neuroevolution.

The rules repository is the central piece that allows introspection of the agent's policy, i. e., behavior, and thus aids its interpretability. It interfaces to the *XRL Rules Extractor*, which takes care of generating rules from the adaptive policy. This happens in two ways: First, given any $(s_t, a_t, \hat{r}_t)$ triplets

procedure $\text{ACT}(s_t\text{: List[SensorReading]}, r_{t-1})$
  $\text{UPDATEREWARD}(rulePolicy, r_{t-1})$
  $\text{UPDATEREWARD}(adaptivePolicy, r_{t-1})$
  $decision_t^{(rule)} \leftarrow \text{DECIDE}(rulePolicy, s_t)$
  $decision_t^{(adaptive)} \leftarrow \text{DECIDE}(adaptivePolicy, s_t)$
  if $\neg decision_t^{(rule)}$ **then**
    $a_t \leftarrow decision_t^{(adaptive)}.a$
  end if
  if $decision_t^{(rule)}$ **then**
    $a_t \leftarrow decision_t^{(rule)}.a$
    if $decision_t^{(rule)}.r < decision_t^{(adaptive)}.r \land decision_t^{(rule)}.a \approx decision_t^{(adaptive)}.a$ **then**
      $\text{STORE}(replayBuffer, decision_t^{(rule)})$
    end if
    if $decision_t^{(rule)}.r < decision_t^{(adaptive)}.r \land decision_t^{(rule)}.a \neq decision_t^{(adaptive)}.a$ **then**
      $a_t \leftarrow decision_t^{(adaptive)}.a$
    end if
  end if
  $\text{ONLINETRAINANDEVOLVE}(adaptivePolicy)$
  $rule \leftarrow \text{TORULE}(adaptivePolicy, decision_t^{(adaptive)})$
  $\text{STORE}(rulesRepo, rule)$
  return a_t
end procedure

Figure 2. Agent Act Routine

from the adaptive policy, a corresponding rule is created by treating the adaptive policy as a black box. Furthermore, the policy ANN is converted into a decision tree [23] and this decision tree is examined for yet-uncovered rules. Thus, the rules extractor constantly feeds rules to the repository, which are ranked according to $(s, a_1, r_1) > (s, a_2, r_2) : r_1 > r_2$.

This completes a two-way interface between the adaptive and the rule-based policy. The rule-based policy yields most of the agent behavior in *learned* situations, allowing to give guarantees with regards to the agent's actions and strategy. The adaptive policy learns to cope with yet unknown situations. At the same time, behavior learned by the adaptive policy is immediately converted into rules, allowing for introspection and extending the behavior governed by guarantees. The rule-based policy is then also able to solve behavioral conflicts. Additionally, rules can be obsoleted by better strategies found by the adaptive policy, which is why sensor readings are always fed to both policies. The rule monitor identifies these obsoleted rules and removes them from the rules repository.

Figure 2 cover the description in this section.

III. Discussion

Obviously, the proposition of the ARL agent architecture is a bold one. Therefore, careful experimentation with benchmark scenarios must be conducted in order to verify the hypothesis underlying the architecture.

Scenario 1 considers voltage regulation as a basic use case. The agent should be able to learn to keep the voltage close to $1.0\,\mathrm{pu}$ in a medium voltage grid, such as the CIGRÉ MV grid. The proposition should hold under time series (i. e., time series data for Photovoltaic (PV) and wind power feed-in, as well as time-series-based customer consumption), as well as under grid constraints (i. e., grid codes). In order to master scenario 1, the ARL agent should be able to cope with the given situations at least as well as, or better, than simple Volt/VAr controllers, as well as simple DRL agents. We expect the rules repository to contain rules similar to that of simple reactive power controllers.

In *Scenario 2*, the agent must cope successfully with changing marginal distributions, such as the introduction of Virtual Power Plants (VPPs) or changing customer behavior. This tests the online learning capabilities of the design. The task is still keeping the voltage close to $1.0\,\mathrm{pu}$. Again, the rules repository will serve as an indicator for the quality of strategies learned, accompanied by the usual DRL metrics, such as reward, objective value, and entropy.

In *Scenario 3*, the ARL agent must succeed against a simple attacker, such as the oscillating attacker by Ju *et al.* [27] or other documented forms of attack. If the ARL agent really constitutes a better concept than the pure DRL approach, then it will not just be able to counter the attack, it will also succeed against different attack strategies. In practice, this is not only another test for the agent's online learning capabilities, but also a way to extract real resilience strategies.

In *Scenario 4*, the ARL agents compete against each other ("attacker" versus "defender"). This is more than just the logical extension of scenario 3: As documented, this forces the agents to sample the extreme areas of the action distribution [12], given a plethora of extractable strategies and documentation of weaknesses of a grid design.

The test scenarios are intended to test the overall behavior of the system: Its ability to adapt through learning, stay interpretable, and give guarantees. Since the research gap addressed by the ARL agent architecture is the combination of learning agent and guaranteed behavior, these scenarios can test the agent by formulating invariants based on expected guarantees.

If successful, we expect the ARL agent to be viable for introduction in grid operator control centers. An initial use case will that of a support and recommender system that helps grid operators to keep situational awareness in complex situations. Later on, the agent can manage parts of the grid (e. g., LV branches with a high number of prosumers) in order to redtce the complexity of grid management. Furthermore, we design the agent to act as a defender against actual cyber attacks.

IV. Conclusion and Future Work

In this paper, we have proposed an agent architecture that fuses rule-based behavior with the learning capabilities of DRL. We did so in order to provide a learning agent that can still give guarantees about its behavior.

In the future, we will develop the respective modules and provide benchmarks and test results, with a special focus on the applicability in CNIs.

ACKNOWLEDGEMENTS

This work was funded by the German Federal Ministry for Education and Research (BMBF) under Grant No. 01IS22071.

REFERENCES

[1] E. M. Veith, *Universal Smart Grid Agent for Distributed Power Generation Management*. Logos Verlag Berlin GmbH, 2017.

[2] E. Frost, E. M. Veith, and L. Fischer, "Robust and deterministic scheduling of power grid actors," in *7th International Conference on Control, Decision and Information Technologies (CoDIT)*, IEEE, IEEE, 2020, pp. 1–6.

[3] M. Sonnenschein and C. Hinrichs, "A distributed combinatorial optimisation heuristic for the scheduling of energy resources represented by self-interested agents," *International Journal of Bio-Inspired Computation*, 2017, ISSN: 1758-0366. DOI: 10.1504/IJBIC.2017.10004322.

[4] O. P. Mahela *et al.*, "Comprehensive overview of multi-agent systems for controlling smart grids," *CSEE Journal of Power and Energy Systems*, vol. 8, no. 1, pp. 115–131, Jan. 2022, Conference Name: CSEE Journal of Power and Energy Systems, ISSN: 2096-0042. DOI: 10.17775/CSEEJPES.2020.03390.

[5] *Flexibility management and provision of balancing services with battery-electricautomated guided vehicles in the Hamburg container terminal Altenwerder*, Energy Informatics, SpringerOpen, 2020. DOI: https://doi.org/10.1186/s42162-020-00129-1.

[6] J. Styczynski and N. Beach-Westmoreland, "When the lights went out: Ukraine cybersecurity threat briefing," *Booz Allen Hamilton*, vol. 12, p. 20, 2016.

[7] A. Aflaki, M. Gitizadeh, R. Razavi-Far, V. Palade, and A. A. Ghasemi, "A hybrid framework for detecting and eliminating cyber-attacks in power grids," *Energies*, vol. 14, no. 18, p. 5823, Jan. 2021, Number: 18 Publisher: Multidisciplinary Digital Publishing Institute, ISSN: 1996-1073. DOI: 10.3390/en14185823. [Online]. Available: https://www.mdpi.com/1996-1073/14/18/5823 (visited on 12/11/2022).

[8] T. Wolgast, E. M. Veith, and A. Nieße, "Towards reinforcement learning for vulnerability analysis in power-economic systems," in *DACH+ Energy Informatics 2021: The 10th DACH+ Conference on Energy Informatics*, Freiburg, Germany, Sep. 2021.

[9] E. M. Veith, N. Wenninghoff, S. Balduin, T. Wolgast, and S. Lehnhoff, *Learning to attack powergrids with ders*, 2022. DOI: 10.48550/ARXIV.2204.11352. [Online]. Available: https://arxiv.org/abs/2204.11352.

[10] O. Hanseth and C. Ciborra, *Risk, complexity and ICT*. Cheltenham, UK: Edward Elgar Publishing, 2007.

[11] R. Diao *et al.*, "Autonomous voltage control for grid operation using deep reinforcement learning," in *2019 IEEE Power & Energy Society General Meeting (PESGM)*, Atlanta, GA, USA: IEEE, Aug. 2019, pp. 1–5, ISBN: 978-1-72811-981-6. DOI: 10.1109/PESGM40551.2019.8973924.

[12] L. Fischer, J. M. Memmen, E. M. Veith, and M. Tröschel, "Adversarial resilience learning—towards systemic vulnerability analysis for large and complex systems," in *ENERGY 2019, The Ninth International Conference on Smart Grids, Green Communications and IT Energy-aware Technologies*, Athens, Greece: IARIA XPS Press, 2019, pp. 24–32, ISBN: 978-1-61208-713-9.

[13] E. Veith, L. Fischer, M. Tröschel, and A. Nieße, "Analyzing cyber-physical systems from the perspective of artificial intelligence," in *Proceedings of the 2019 International Conference on Artificial Intelligence, Robotics and Control*, ACM, Dec. 2019, ISBN: 978-1-4503-7671-6.

[14] J. Schrittwieser *et al.*, "Mastering Atari, Go, Chess and Shogi by planning with a learned model," pp. 1–21, 2019. arXiv: 1911.08265. [Online]. Available: http://arxiv.org/abs/1911.08265.

[15] S. Fujimoto, H. van Hoof, and D. Meger, "Addressing function approximation error in actor-critic methods," *arXiv:1802.09477 [cs, stat]*, Oct. 22, 2018. arXiv: 1802.09477. [Online]. Available: http://arxiv.org/abs/1802.09477 (visited on 12/22/2021).

[16] J. Schulman, F. Wolski, P. Dhariwal, A. Radford, and O. Klimov, "Proximal policy optimization algorithms," Jul. 19, 2017. arXiv: 1707.06347. [Online]. Available: http://arxiv.org/abs/1707.06347.

[17] T. Haarnoja, A. Zhou, P. Abbeel, and S. Levine, "Soft actor-critic: Off-policy maximum entropy deep reinforcement learning with a stochastic actor," *arXiv:1801.01290 [cs, stat]*, Aug. 8, 2018. arXiv: 1801.01290. [Online]. Available: http://arxiv.org/abs/1801.01290 (visited on 12/22/2021).

[18] F. P. Such *et al.*, "Deep neuroevolution: Genetic algorithms are a competitive alternative for training deep neural networks for reinforcement learning," 2017. arXiv: 1712.06567. [Online]. Available: http://arxiv.org/abs/1712.06567.

[19] M. I. Radaideh *et al.*, *NEORL: NeuroEvolution optimization with reinforcement learning*, Dec. 1, 2021. DOI: 10.48550/arXiv.2112.07057. arXiv: 2112.07057[cs]. [Online]. Available: http://arxiv.org/abs/2112.07057 (visited on 12/11/2022).

[20] R. Cheng, G. Orosz, R. M. Murray, and J. W. Burdick, "End-to-end safe reinforcement learning through barrier functions for safety-critical continuous control tasks," *33rd AAAI Conference on Artificial Intelligence, AAAI 2019, 31st Innovative Applications of Artificial Intelligence Conference, IAAI 2019 and the 9th AAAI Symposium on Educational Advances in Artificial Intelligence, EAAI 2019*, pp. 3387–3395, 2019, ISSN: 2159-5399. DOI: 10.1609/aaai.v33i01.33013387. arXiv: 1903.08792.

[21] K. Khetarpal, M. Riemer, I. Rish, and D. Precup, *Towards continual reinforcement learning: A review and perspectives*, Dec. 24, 2020. DOI: 10.48550/arXiv.2012.13490. arXiv: 2012.13490[cs]. [Online]. Available: http://arxiv.org/abs/2012.13490 (visited on 10/14/2022).

[22] E. Puiutta and E. M. S. P. Veith, "Explainable reinforcement learning: A survey," in *Machine Learning and Knowledge Extraction. CD-MAKE 2020*, Dublin, Ireland: Springer, Cham, 2020, pp. 77–95. DOI: 10.1007/978-3-030-57321-8_5.

[23] C. Aytekin, *Neural networks are decision trees*, Oct. 25, 2022. DOI: 10.48550/arXiv.2210.05189. arXiv: 2210.05189[cs]. [Online]. Available: http://arxiv.org/abs/2210.05189 (visited on 11/09/2022).

[24] G. Brockman *et al.*, *OpenAI gym*, 2016. eprint: arXiv:1606.01540.

[25] E. M. Veith *et al.*, "Analyzing power grid, ICT, and market without domain knowledge using distributed artificial intelligence," in *CYBER 2020, The Fifth International Conference on Cyber-Technologies and Cyber-Systems*, IARIA, IARIA, 2020, pp. 86–93.

[26] C. Posthoff and B. Steinbach, "Extremely complex problems: Binary models for computer science," in Jan. 2019, pp. 339–360, ISBN: 978-3-030-02419-2. DOI: 10.1007/978-3-030-02420-8_8.

[27] P. Ju and X. Lin, "Adversarial attacks to distributed voltage control in power distribution networks with DERs," in *Proceedings of the Ninth International Conference on Future Energy Systems*, ACM, 2018, pp. 291–302, ISBN: 9781450357678. DOI: 10.1145/3208903.3208912.

Development of Occupants' Behavior Model in Urban Scale Using Dynamic Time Warping and Particle Swarm Optimization Algorithms Based on National Lifetime Survey

Hengxuan Wang
Graduates School of Human-Environment Studies
Kyushu University
Fukuoka, Japan
e-mail: whx7174615@gmail.com

Daisuke Sumiyoshi
Faculty of Human-Environment Studies
Kyushu University
Fukuoka, Japan
e-mail: sumiyoshi@arch.kyushu-u.ac.jp

Abstract—**For the target of energy demand estimation of residential buildings in urban scale, occupants' behavior model has been paid much attention. In this paper, a new model for simulating occupants' behavior schedules in urban scale has been proposed using only public stochastic data (national lifetime survey) combined with Dynamic Time Warping and Particle Swarm Optimization algorithms. We use this proposed model to simulate the working-male's behavior schedules with 5-mintues interval in resting day as an example. The simulated results - percentages of occupants adopt the given behavior at specific moments are calculated and compared with public stochastic data to verify the accuracy. Compared with existing models, the proposed model is more efficient and accurate. We believe this model could be useful for building energy demand estimation in urban scale combined with appliance operation possibility based on occupants' behaviors.**

Keywords-occupants' behavior model; energy demand of residential buildings; public stochastic data; particle swarm optimization; dynamic time warping.

I. INTRODUCTION

A. Background

To get the target of decarbonized society in 2050 [1], the Japanese government is promoting the introduction of decentralized renewable energy devices in urban area to reduce carbon emissions. But without a suitable introduction plan, the surplus electricity generated from excessive devices would disturb the balance between the supply and demand of power system or failing to get decarbonization target because of insufficient devices. Therefore, in building sector, it is essential to develop the decentralized energy introduction plan based on the energy demand of buildings in urban area.

The renewable energy is limited to natural condition (e.g., solar energy) and outpower changes dramatically over time. Therefore, the energy demand of buildings should be estimated with high temporal resolution. Non-residential buildings (e.g., office) have temporal characteristics of energy demand because of fixed schedule of users. However, the energy demand of residential building is decided by appliances' operation, which is influenced by the behavior of the occupants with significantly personal characteristics. In previous studies about energy demand estimation for residential buildings, the behavior schedules of occupants had been set to several cases. This assumption would significantly affect the accuracy of results. The reason is that even the same type occupants in urban scale would have numerous kind behaviors at the same time, but there are only a few cases in these few schedules that would overlay the peak or trough energy demand amount. Thereby, the demand results and the amount of renewable energy devices need to be introduced would be a departure from reality. Thus, a method to simulate the occupants' behavior schedules in urban scale is very essential for the plan of introduction of decentralized renewable energy in urban scale.

B. Related Work

There is much previous research about the occupants' behavior model in urban scale [2]. The models could be divided into two types based on whether to use dataset called Time Use Data (*TUD*), which describe occupants' behavior by time.

For the first type without *TUD*, in [3], a model was developed using only public stochastic data of *TUD* include mean and standard deviation of behaviors' duration time in a day and percentages of occupants adopt the behavior at special moments by 15-mintues interval of a day. They firstly selected the behaviors according to probabilities and arranged their total duration time into 24 hours. Next, they placed the first behavior into the slot in timeline according to random number and placed the next behavior into the end of previous behavior one by one. As one merit, this method could generate the occupants' behavior schedules with only public stochastic data. But the accuracy of the simulation results was greatly influenced by the first behavior's inserted slot, which was decided randomly. Also, the results had not been validated.

For the type of models using *TUD*, in [4], they developed a model using the Markov Chain, which is a stochastic model to determine the transition of behavior from another only depend on the condition at the previous time step. They collected the *TUD* from a great number of households and analysis the transition probability between behaviors. But the behavior items were limited in at room or not.

In [5], they proposed a Markov Chain model and expanded analysis of the number of behavior items. They simulated the household's members independently. These Markov Chain models considered and simulated the transitions probability between the behaviors precisely, but the accuracy of behavior duration time was dependent on the timing and number of behavior transitions. This could be a weakness for simulating the occupants' behavior schedules. In [6], they developed a occupants' behavior model dealing with above problems. They divided behaviors into routine and non-routine and considered them separately. The behaviors' duration time and transition probabilities between them were acquired by analyzing the *TUD* from the national time-use survey conducted by Statistics Japan in 2006 and been utilized for placing the behaviors into the timeline. They firstly placed the routine behaviors (including sleeping, commuting to work & school, dining and bathing) into timeline, then selected the non-routine behaviors according to the probabilities and placed them in the gap between the routine behaviors until all gaps had been filled. They improved the model in [7] considering the interaction among household members (e.g., household members always have dining together at one time and bathing one by one) and time-dependent characteristics of the specific behaviors (e.g., for a single person, the personal washing often happens immediately after waking up or breakfast, but it's not shown in *TUD* because it was originated from a wide range of people.). In [8], they explored several machine learning methods to pre-process the *TUD* to improve the accuracy of behavior model. Although the duration time and transition probabilities of behaviors were detailed considered in their model, predetermining the number of behavior occurrences with a subjective assumption was made. (e.g., three meals over a day, one sleeping in the evening with long period), but according to the public stochastic data, there are also sleeping at the daytime for many type people), this might be a weakness of their model for ignoring this specifical cases. On the other hand, raw *TUD* are required to make this kind of model while only public stochastic data are available in many countries.

As mentioned above, until now there are many developed occupants' behavior models in urban scale with own strengths and weaknesses. But there is still no precise occupants' behavior model that do not require prior analysis of large amount of raw *TUD*, pre-classification of behaviors according to routinely or not and predetermined number of occurrences with subjective assumption.

C. Purpose of the paper

In this paper, a new occupants' behavior model using only public stochastic data without raw *TUD* has been proposed. Compared with existing models in previous research, this model could ensure the accuracy and efficiency.

Section II introduces the detailed procedures of the proposed model. Section III corrects the simulation processes based on the simulation results. In Section IV, the final simulation result of working-male in resting day is shown. In Section V, the conclusions about model's features and weaknesses are introduced. Based on that, the directions of improving model in future are also introduced.

II. Proposed Behavior Model

A. Parameters of purposed occupants' behavior model

The proposed model simulates the occupants' behavior schedules with 5-min interval based on the public stochastic data called National Lifetime Survey in 2020 from Japan Broadcast Institution (*NHK*) [9]. It should be noted that target of simulating behavior schedules is to estimate energy demand of residential building, so the behaviors that have no relationship with energy demand in residential building (e.g., working outside, commuting to work, school) have not been simulated in this paper. This assumption, which is one of the differences between the previous studies, can greatly simplify the model. Table I shows the classification result of behaviors on public stochastic data. These behaviors have been simplified into 24 types and divided into interior and exterior. According to whether using appliances, the interior behaviors are further divided into two categories.

To make the model, the public stochastic data would be utilized include:

- ***PM***: probabilities of adopting given behavior by 15-minutes interval (the data has been processed into 5-minutes interval by liner interpolation).
- ***PA***: probabilities of adopting given behavior over a day.
- ***MTB***: average duration time of adopting given behavior.
- ***SDTB***: standard deviation of duration time of given behavior.

Some samples of public stochastic data are shown in Table II.

During the process of simulation, a blank timeline with 288 time slots (time of a day with 5-minutes interval) is generated firstly and prepares for filling up with behaviors

TABLE I. CLASSIFICATION OF BEHAVIORS

Interior Behavior (lighting & HVAC used)		**Exterior Behavior**
Appliance used	***Non-appliance used***	
eating	children care	shopping
washing	leisure	conversation personal relationships
sleeping	reading newspaper	work
hobbies, entertainment and culture (with Internet)	reading magazines comics	leisure and exercise
hobbies, entertainment and culture (without Internet)		class and lecture
cooking, cleaning, laundry		commuting
radio		sporting
household chores		

TABLE II. SAMPLE OF PUBILC STOCHASTIC DATA OF WORK-MALE IN SUNDAY AND TARGET BEHAVIOR -SLEEPING

Behavior	***PA***	***MTB***	***SDTB***
sleeping	99.20%	8:25	2:07
eating	97.60%	1:38	0:52
washing	96.00%	1:04	0:34

Time	***PM***
0:00	70.20%
0:05	71.27%
0:10	72.33%

separately. The given behavior's occurrences will span corresponding time slots in the timeline depend on its duration time length. The detail steps are explained as:

1) Iterates over all given behaviors in order of the ***PA***'s values and determines whether to adopt based on its ***PA***.
2) Once the given behavior has been adopted, the duration time (***TB***) is determined according to the Gaussian Distribution defined by ***MTB*** and ***SDTB***.
3) To insert these given behaviors into the timeline, it is critical to decide several parameters' solution of the given behavior including:
   a) ***n***: number of behavior occurrences. (***n*** =1~4 randomly)
   b) *sm*: start moment of each behavior occurrence. ***SMN***: $[sm_1, sm_2, \cdots sm_n]$
   c) *pn:* probability of each number of behavior occurrences. (e.g., pn_2: probability of behavior occurring twice) ***PNN***: $[pn_1, pn_2, \cdots pn_n]$
   d) *pt*: each occurrence's duration time as percentage of ***TB*** in a large number of schedules. ***PTN:*** $[pt_1, pt_2, \cdots, pt_n]$ (e.g., Figure 1 shows the difference between ***PTN₁***: [1/3,1/3,1/3] and ***PTN₂***: [1/6,1/3,1/2]

B. Start moments of behavior occurances

It is necessary to decide ***SMN*** 's solution to determine the positions of timeline where the behavior occurrences are going to be inserted.

Figure 2 shows the process of deciding the start moment of each behavior occurrence by the cumulative distribution of ***PM***. In detail, one day is divided into ***n*** time regions, which have the same sum of ***PM***. The start moment of splitting time region is generated randomly (e.g., 6:30). It is assumed that object behavior occurs once in each time region. Based on that assumption, the cumulative distribution function of ***PM*** in each time region has been calculated to determine the start moment of each occurrence.

C. Dynamic Time Warping

Different from ***SMN***, it is impossible to get ***PNN&PTN*** solution based on the existing public stochastic data merely.

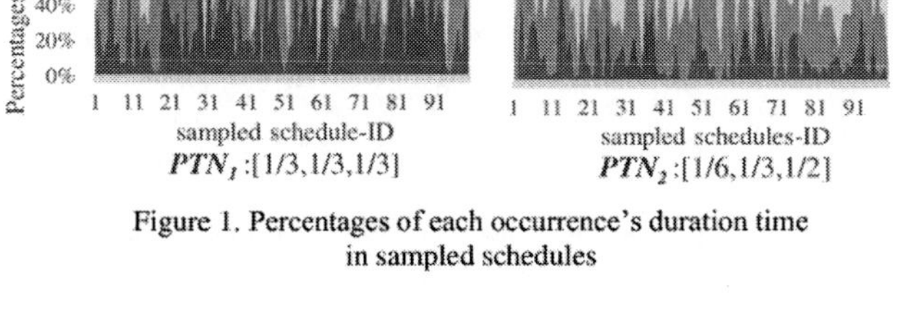

Figure 1. Percentages of each occurrence's duration time in sampled schedules

It is necessary to introduce parameter optimization method to obtain the optimal ***PNN&PTN*** solution.

To verify the fitness of ***PNN&PTN*** candidate solution, we introduce the objective function to compare ***PM*** and probability of adopting given behavior at 5-min interval, which is calculated by schedules generated using ***PNN&PTN*** candidate solution (***PM'***). For ***PM*** and ***PM'*** are both time series data, Dynamic Time Warping (*DTW*) introduced in [10] is used as objective function to measure their similarity. *DTW* of ***PM*** and ***PM'*** is calculated by (1):

$$DTW(\boldsymbol{PM}, \boldsymbol{PM'}) = \min \sqrt{\sum (x_i - y_j)^2} \quad (i,j) \in L \qquad (1)$$

$$\boldsymbol{PM} = [x_1, x_2, \cdots, x_{287}],\ \boldsymbol{PM'} = [y_1, y_2, \cdots, y_{287}]$$

The list of index pairs $L = [l_0, l_1, \cdots, l_{287}]$ shows the matching pairs of the elements of ***PM*** and ***PM'*** (e.g., $l_k = (i_k, j_k)$ shows the x_{i_k} and y_{j_k} would be matched) that satisfies the following properties are shown in (2) (3) (4):

$$0 \le i_k, j_k \le 287 \qquad (2)$$

$$l_0 = (0,0),\ l_{287} = (287, 287) \qquad (3)$$

$$l = (i_k - 1, j_k)\ or\ (i_k, j_k - 1)\ or\ (i_k - 1, j_k - 1) \qquad (4)$$

Different from the traditional matching method, which would match ***PM*** and ***PM'*** at the same index pairs ((x_1, y_1), (x_2, y_2), ..., (x_{287}, y_{287})). In *DTW*, based on the above properties, there is a large number (*T*) of possible matching solutions as candidates, which is shown in (5):

$$T[l_0, l_1, \cdots, l_{287}] = \begin{bmatrix} (x_0, y_0) & (x_1, y_0) & (x_1, y_1) & \cdots & (x_{287}, y_{287}) \\ \vdots & \vdots & \vdots & \ddots & \vdots \\ (x_0, y_0) & (x_0, y_1) & (x_0, y_2) & \cdots & (x_{287}, y_{287}) \end{bmatrix} \qquad (5)$$

all matching solutions' distances between ***PM*** and ***PM'*** would be compared and the smallest one would be called *DTW*. By calculating the *DTW* obtained from different ***PNN&PTN*** candidate solutions, the most suitable

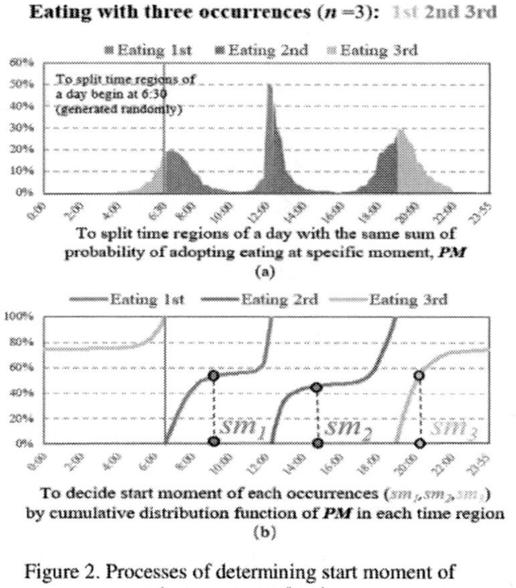

Figure 2. Processes of determining start moment of each occurrence of eating

PNN&PTN solution would be decided with the minimal *DTW*.

D. Patricle Swarm Optimization

As mentioned above, the best ***PNN&PTN*** solution can be found by finding the minimal *DTW*. In this paper, we use Particle Swarm Optimization (*PSO*) algorithm to find the minimal *DTW*. *PSO* is an evolutionary algorithm introduced in [11] that could optimize a problem by iteratively trying to improve a candidate parameters' solution to get the better position in a D-dimensional space (D is the number of parameters).

In the process of the *PSO* algorithm, firstly, a large number of particles have been generated and each particle is a candidate solution of ***PNN&PTN*** with different *DTW* result. At 1st iteration, particle' initial position (p_1) and velocity (v_1) are randomly generated. p_1 means ***PNN&PTN*** solution and v_1 means the distance between p_1 and p_2 (position at 2nd iteration) as showed in (6). These particles make up a cloud that covers the entire space, then the *DTW* of all particles are calculated to decide their fitness. Based on fitness values, the globally best particle position (pg_1) and locally best particle position (pl_1) are determined. As showed in (7), according to pg_1, pl_1 and p_1, v_1 would be updated to v_2, which would continue to update p_2 to p_3. With the iteration advancing, the cloud contracts gradually and performs the exploration for best ***PNN&PTN*** solution with minimal *DTW*.

$$p_{k+1} = p_k + v_k \tag{6}$$

$$v_{k+1} = wv_k + \varphi_1(pg_k - p_k) + \varphi_2(pl_k + p_k) \tag{7}$$

k :	k^{th} iteration	pl_k :	locally best particle's position at k^{th} iteration
w :	inertia weight	φ_1, φ_2 :	$\varphi_1 = c_1 r_1$, $\varphi_2 = c_2 r_2$
v_k :	particle's velocity at k^{th} iteration	r_1, r_2 :	random numbers in the range [0,1]
p_k :	particle's position at k^{th} iteration	c_1, c_2 :	$c_1 = c_2 = 2$
pg_k :	globally best particle's position at k^{th} iteration		

E. Process of proposed model

Figure 3 shows the proposed model's all processes for simulating the behavior schedules. Using this model, 1000 behavior schedules have been generated and evaluated.

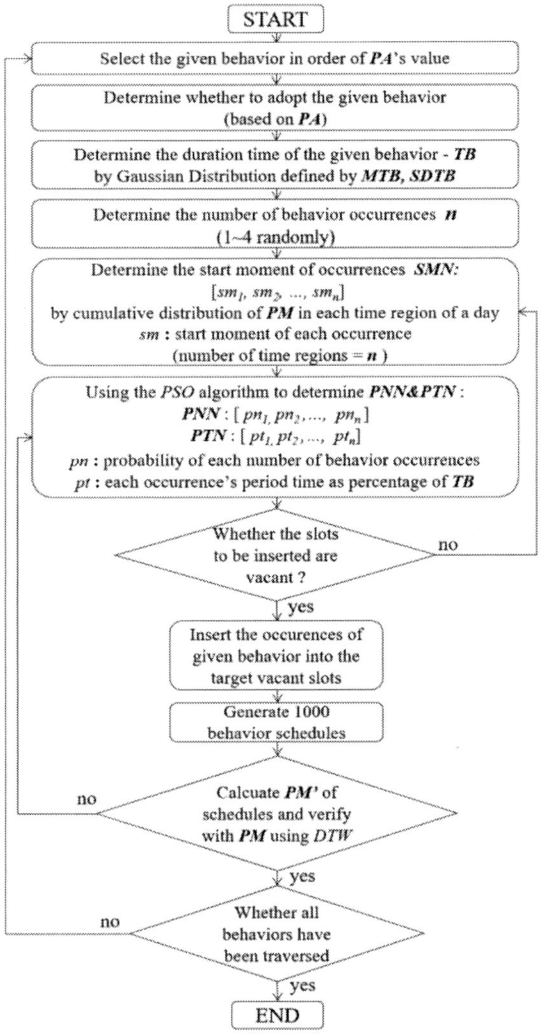

Figure 3. Simulation processes of proposed model

III. CORRECTION OF SIMULATION PROCESS

According to the schedule results, there are two significant errors include:

1) Figure 4 shows the results of sleeping are inaccurate thoroughly.
2) Figure 5 shows the delaying of start moments of ***PM'*** compared with ***PM***.

The above errors would be dealt with as follows:

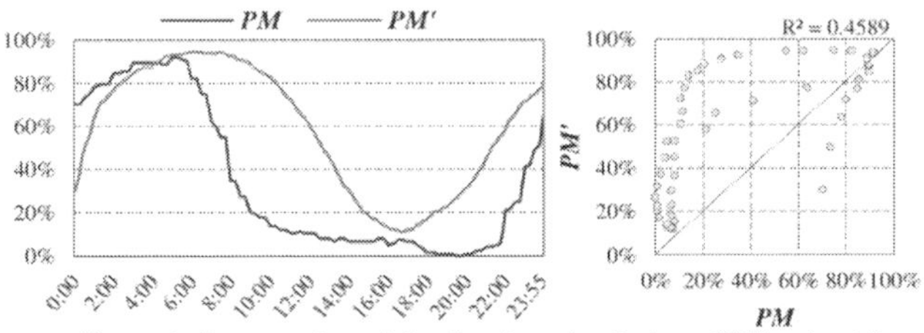

Figure 4. Comparation of simulated stochastic data, ***PM'*** and public stochastic data, ***PM*** (sleeping)

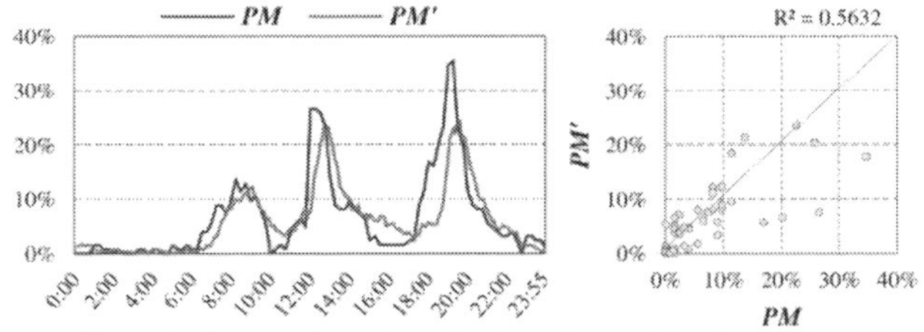

Figure 5. Comparation of simulated stochastic data, ***PM'*** and public stochastic data, ***PM*** (eating)

A. Correction of sleep simulation process

The error 1) can be attributed to the inaccurate determination of start moments of sleeping. Different from other behaviors, people always have a long period sleeping in the evening and add a short period sleeping during the daytime. For this type of behavior with a clear temporal characteristic, the method deciding behavior start moments based on sum of ***PM*** is not suitable any longer. To solve this problem, we revise the model to simulate sleeping behavior in the following process:

a) The number of sleeping occurrence (***n***) is set to 1~2. If sleeping occurs once, it occurs at night; if sleep occurs twice, the first and longer sleeping occurs in the evening and second one occurs at daytime.

b) For sleeping in the evening, people wake up at a more concentrated time than when they fall asleep. Therefore, we use the moments of waking up (ending of sleeping) to decide the position of sleeping in timeline where being inserted into.

c) The range of end moments of first sleeping in the evening is set as 0:00-12:00, the range of start moments of second sleep is set as 12:00-18:00. The specific moments in the range are searched by *PSO* method too.

To sum up, the parameters of sleeping for *PSO* method are reset showed as (8):

$$\boldsymbol{SMN} = [sm_1, sm_2]\ \ \boldsymbol{PNN} = [pn_1, pn_2]\ \ \boldsymbol{PTN} = [pt_1, pt_2] \quad (8)$$

After the calculation by *PSO*, Figure 6 shows the results of sleeping's ***PM*** and ***PM'*** by this revised process, which is better than original one.

B. Correction of start moment

For the error 2), the reason being considered is that the decision of start moment based on cumulative distribution function of ***PM*** always drop behind actual situation. To solve this issue, the new parameter *ad* is introduced to adjust the ***SMN***: $[sm_1 + ad, sm_2 + ad, \cdots, sm_n + ad]$. The decision of *ad* is also calculated by *PSO* method. Therefore, the solution of ***PNN&PTN***, *ad* would be decided together by minimal *DTW*. Figure 7 shows the simulation results of ***PM*** and ***PM'*** after the adjustment of behavior start moments and it demonstrates higher accuracy than before.

IV. SIMULATION RESULTS

Table III shows the calculation results of behavior washing's parameter solutions by *PSO* algorithm of working-male in resting day. In Figure 8, the probability distribution of all target behaviors is shown. The result shows that ***PM'*** agreed well with ***PM*** and it confirms our model's accuracy.

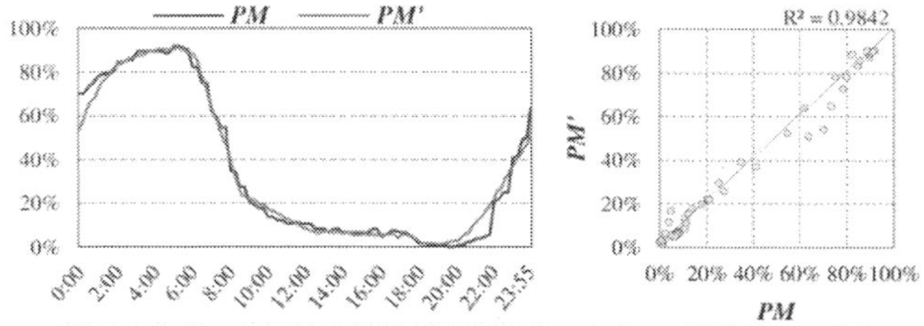

Figure 6. Comparation of simulated stochastic data, ***PM'*** after revising and public stochastic data, ***PM*** (sleeping)

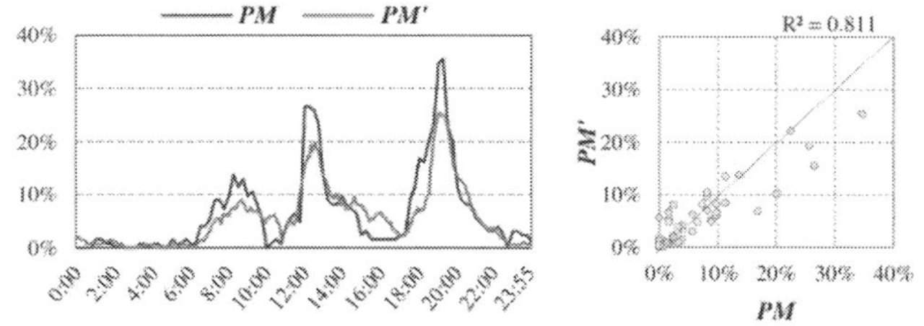

Figure 7. Comparation of simulated stochastic data, ***PM'*** after revising and public stochastic data, ***PM*** (eating)

TABLE III. PARAMETER SOLUTIONS OF BEHAVIOR WASHING CALCULATED BY *PSO* ALGORITHM

Behavior	*Parameter Solutions – PNN, PTN, ad* $[pn_1, pn_2, pn_3, pn_4]$ $[pt_1, pt_2, pt_3, pt_4]$ $[ad]$
washing	***PNN*** = [0.9%, 18.7%, 31.8%, 48.6%] ***PTN*** = [21.7%, 20.6%, 18.2%, 39.5%] *ad* = [3]

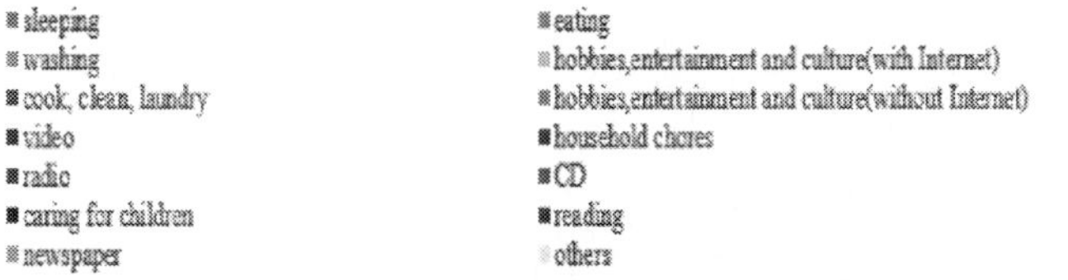

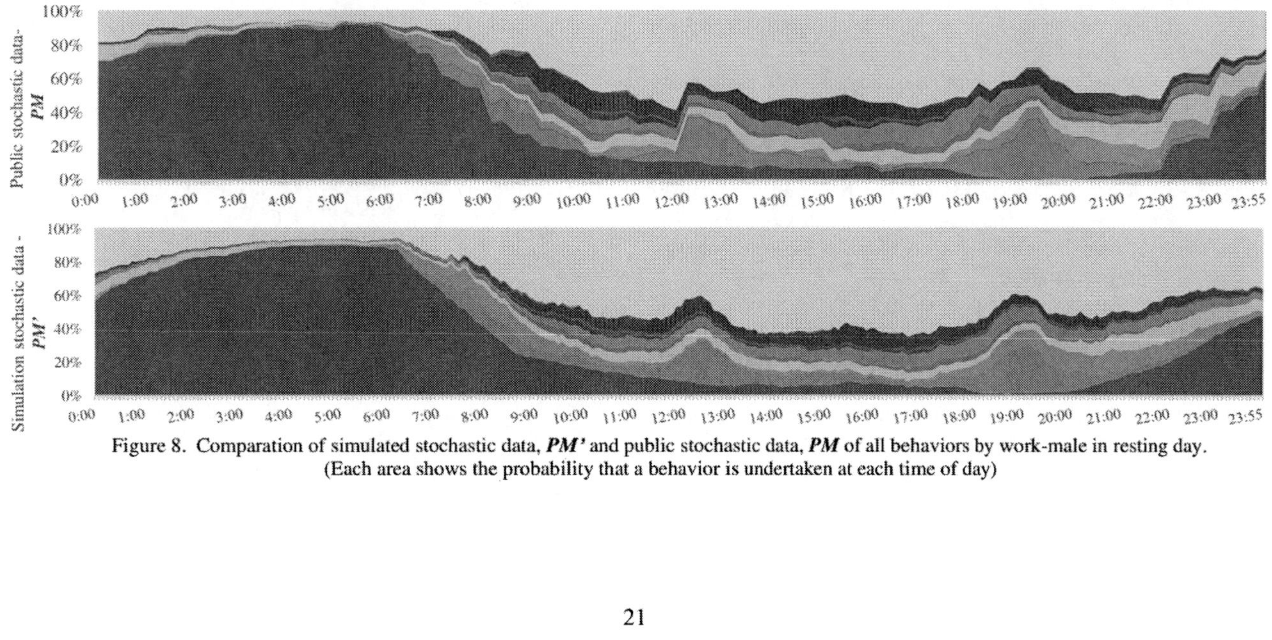

Figure 8. Comparation of simulated stochastic data, ***PM'*** and public stochastic data, ***PM*** of all behaviors by work-male in resting day. (Each area shows the probability that a behavior is undertaken at each time of day)

V. CONCLUSION AND FUTURE WORK

This paper proposes a model based on public stochastic data to generate occupants' behavior schedules at home. It could be used in energy demand estimation of residential buildings in urban scale. More than just simulation for behavior schedules based on past public stochastic data, it might be useful for assuming the future people behavior change by altering the inputting public stochastic data (e.g., the working time at home increased because of covid-19). Compared with existing behavior model's research, the proposed model has the following features:

- Generating occupants' behavior schedules based on public stochastic data only. Without statistical analysis of large amounts of raw *TUD*, which is not available in many countries, making the behavior model simpler and more efficient.
- No classifying the behaviors or setting the specific number and duration time of behavior occurrences. This feature could exclude the errors from subjective assumptions.
- Utilizing the *PSO* and *DTW* algorithm to search the suitable number of behavior occurrences and percentages of occurrences' duration time. It would make the simulation results match the public stochastic data as closely as possible.
- Deciding the start moments of behavior based on cumulative distribution of public stochastic data. As the simulation results do not agree with the public stochastic data, the start moments calculated by the above method have been corrected using *PSO* algorithm.

It should be noted that by *PSO* algorithm merely, the ***SMN*** could be determined without using cumulative distribution of ***PM***. But with the assistance of cumulative distribution, the *PSO* algorithm could narrow the search range and get solution quickly. Using this revised model, the working-male's behavior schedules in resting day have been generated. The result shows that our model has a good accuracy. But there are also several drawbacks:

a) During some time interval s (e.g., at 12:00~12:30 and 22:00~22:30, the public stochastic data - ***PM*** of eating, sleeping increase rapidly, but the simulation results - ***PM'*** fail to reflect such phenomenon.

b) No consideration of interaction between behaviors (e.g., people are likely to wash themselves when they wake up, but behavior transition between sleeping and washing can't be simulated in a single schedule).

c) Fail to consider the interaction between the family members (e.g., having a meal together). This interaction is import for residential building energy demand estimation.

In future, we are going to deal with these drawbacks to improve the behavior model. About drawback *a)*, more in depth analysis of public stochastic data especially in specific time interval will be done so that different weights will be given during these time intervals in simulation process. For *b)*, which had been raised in much previous research, analysis the raw *TUD* to get the results of behavior transition probabilities is a feasible option. And about last *c)* as mentioned in [7], when several kinds of schedules from different people in a family are required for residential building's electricity demand estimation, it could be a solution to choose schedules from generated schedule database, which have meals at the same time and bathing in sequence. Also, it should be noted that the behavior schedules could not be used directly in the energy demand estimation model without appliance operation possibility based on behaviors. More work about the relationship between behavior and appliance operation is also necessary in future.

REFERENCES

[1] Ministry of the Environment, Japan, Outlines of Japan's Long-term Strategy under the Paris Agreement, 2019.

[2] G. Happle, J. A. Fonseca, and A. Schlueter, "A review on occupant behavior in urban building energy models," Energy & Buildings, vol. 174, pp. 276-292, Sept. 2018, doi: 10.1016/j.enbuild.2018.06.030.

[3] J. Tanimoto, A. Hagishima, and H. Sagara, "Validation of methodology for utility demand prediction considering actual variations in inhabitant behavior schedules," Journal of Building Performance Simulation, vol. 1, no. 1, pp. 31-42, Mar. 2008, doi:10.1080/19401490701868471.

[4] I. Richardson, M. Thomson, and D. Infield, "A high-resolution domestic building occupancy model for energy demand simulations," Energy and Building, vol. 40, no. 8, pp. 1560-1566, Feb. 2018, doi:10.1016/j.enbuild.2008.02.006.

[5] J. Widén, et al., "Constructing Load Profiles for Household Electricity and Hot Water from Time-Use Data – Modelling Approach and Validation," Energy and Building, vol. 41, no. 7, pp. 753-768, Mar. 2009, doi:10.1016/j.enbuild.2009.02.013.

[6] Y. Yamaguchi and Y. Shimoda, "Evaluation of behavior model of occupants in home based on Japanese national time use survey," Proceedings of Building Simulation 2015: 14th Conference of IBPSA, Dec. 2015, pp. 2897-2904, doi: 10.26868/25222708.2015.2520.

[7] Y. Yamaguchi and Y. Shimoda, "A stochastic model to predict occupants' activities at home for community-/urban-scale energy demand modelling," Journal of Building Performance Simulation, vol. 10, no. 5-6, pp. 565-581, Jun. 2017, doi: 10.1080/19401493.2017.1336255.

[8] Y. Li, Y. Yamaguchi, and Y. Shimoda, "Impact of the pre-simulation process of occupant behavior modelling for residential energy demand simulations," Journal of Building Performance Simulation, vol. 15, no. 3, pp. 287-306, Mar. 2021, doi:10.1080/19401493.2021.2022759.

[9] Japan Broadcast Institution (*NHK*), National Lifetime Survey in 2020, Japan.

[10] H. Sakoe and S. Chiba, "Dynamic programming algorithm optimization for spoken word recognition," IEEE transactions on acoustics, speech, and signal processing, vol. 26, no. 1, pp. 43-49, Feb. 1978, doi:10.1109/TASSP.1978.1163055.

[11] J. Kennedy and R. Eberhart, "Particle swarm optimization", Proceedings of ICNN'95 – International Conference on Neural Networks, Nov. 1995, pp. 1942-1948, doi: 10.1109/ICNN.1995.488968.

Machine Learning and Optimisation to Improve Energy Utilisation

Srinath Ramagiri
Brunel University
Kingston Ln, Uxbridge UB8 3PH
London, United Kingdom
e-mail: srinath.ramagiri@brunel.ac.uk

Evelyne El Masri
Brunel University
Kingston Ln, Uxbridge UB8 3PH
London, United Kingdom
e-mail: evelyne.elmasri@brunel.ac.uk

Arman Zonuzi
University of Sheffield
Brunel Way, Catcliffe, Rotherham S60 5WG
United Kingdom
e-mail: arman.zonuzi@namrc.co.uk

Ahmed Teyeb
Brunel University
Kingston Ln, Uxbridge UB8 3PH
London, United Kingdom
e-mail: ahmed.teyeb@brunel.ac.uk

Shehan Lowe
University of Sheffield
Brunel Way, Catcliffe, Rotherham S60 5WG
United Kingdom
e-mail: shehan.lowe@namrc.co.uk

Tat-Hean Gan
Brunel University
Kingston Ln, Uxbridge UB8 3PH
London, United Kingdom
e-mail: tat-hean.gan@brunel.ac.uk

***Abstract*— The world is moving towards a conservative approach to fulfilling its energy needs due to inevitable uncertainty and disruptions in the supply chain. In addition, climate change, the availability of materials, and making them sustainable through recycling are other topics of high interest. Energy is a common item among all the industries, and demand for it keeps increasing due to developmental activities. In this work, we aim to improve the efficiency of utilising the available energy in the material processing industries. Mining the ore, extracting the material of interest, melting the material, and manufacturing the required components are typical processes in these industries. The manufacturing of the components also includes a heat treatment process. For example, the heat treatment process demands 20% of the total energy in a non-ferrous foundry. Pre-heating and heat treatment operations consume a significant amount of energy in the ferrous-based industry. We intend to investigate the processes in these industries and create a machine-learning model of the processes involved. Later, we use the machine learning models to build an optimization framework that provides the optimal process operating parameters to achieve the best output while using the least amount of energy.**

Keywords- machine-learning; Optimisation; heat-treatment; energy-efficiency.

I. INTRODUCTION

Heat treatment processes are an important stage in materials processing in which component properties are modified to suit a particular application. In this process, mechanical and physical properties, such as ductility, hardness, toughness, wear resistance, and strength are changed without changing the designed shape and size of the component [1]. In general, heat treatment processes are carried out to improve strength in the case of loaded members and wear resistance in the case of moving parts, however, it can also be used to improve the machinability, formability of materials. The changes in the properties of the material are made possible thanks to the changes which occur at molecular structure/microstructure level. The structure of the material is a function of two factors; (1) Grain size (2) Grain structure. These two components of microstructure of a material define its mechanical and physical properties. Also, heat treatment process is often coupled with pre and post heating process which enable us to utilize energy effectively besides improving the product performance.

In total, it can be observed that there are several parameters involved in heat treatment process such as, chemical composition of alloy, dimensions and shape of the component to be heat treated, micro structural, physical and mechanical properties, energy required for the heat treatment process, etc. Depending on specific objective, some of the parameters will be input before/during heat treatment and others will be output parameters. Irrespective of our objectives if we end up with more than 4 parameters, which vary then a complex problem needs to be solved to realize the effect of each parameter on the set out put parameters. To address this issue regression models are used. A brief literature review of the regression models used to map input to output parameters in heat treatment process is presented in the following section.

The rest of the paper is organized as, details of regression model and its results are presented in Section III, details of optimization framework and its results are presented in Section IV and the conclusions are presented in Section V.

II. LITERATURE

In the case of metals, Johnson [2], Avrami [3] proposed analytical models. Using their work, numerical simulations were developed to study heat treatment process. The simulation results reduced the effort of extensive experimentations (hence reduction in energy consumption) however they lack accuracy [4]. For example, Maisuradze [5] proved that the simulations predict strength parameter in-accurately in a heat treatment process. To address this issue computer aided simulations are developed which give a better accuracy than the simulations [6] – [9].

In general, an important task in a heat treatment model development is to predict the micro-structural properties accurately as they predominantly influence output quality parameters. In this way it is easy to suggest the initial/input parameters of a heat treatment. In this regard, data driven solutions are proven to accelerate the problem solving [10] [11]. Related works in this field include, Homer et al. [12] and Zhu et al. [13] investigated grain boundaries in polycrystalline material using machine learning tools, Raccuglia et al. [14] created a classification model to predict successful and failed experiments using vast amount of experimental data on materials. Agrawal et al. [15] [16] created a machine learning model which predicts fatigue strength of steel using composition and processing parameters. In [17] regression models are used to predict four mechanical properties after heat treatment process. The authors have used five different regression models of which random forests performed well in predicting the mechanical properties and they developed mathematical expressions out of it.

Heat treatment and other processes of glass also results in change in microstructure and hence mechanical properties. Structural and physical properties of strontium borate glass are evaluated against chemical composition using regression analysis by Masai H et al. [18]. Samuel B. O. et al. [19] have developed an optimization technique using which they modelled glass material for composites of particular flexural strength using Taguchi and general regression. In case of glass, a lot of attention is given to the manufacturing processes such as cutting, grinding, etc. In [20] [21], the authors have created a Neural Network (NN) model to predict material removal rate and surface roughness parameters in a laser machining process. Shimaa et al. [22] created a machine learning model of abrasive jet machining of glass in which material removal rate is correlated with process variables of machining process. Bezzera et al. [23] developed a machine learning model using NN to predict shear stress-strain behavior of CFRP material.

There is a lot of research literature in which investigations are carried out on implementing optimization methods with regression models as basis for prediction. Although regression models developed on machine learning techniques perform as per the expectations, the developed models suffer from a dis-advantage. The solution space of the machine learning models may refer to a local maxima or minima, over/under fitting and slow convergence. These dis-advantages can be identified and overcome by implementing an optimization technique to rigorously search the solution space for a specific solution that fits our application [24] [25] [26]. In recent research works, deep learning models such as Deep Neural Networks (DNN), Convolutional Neural Networks (CNN), and Recurrent Neural Networks (RNN) are used for forecasting energy demand forecasting. Works of [27] [28] [29] have shown that using more than one type of deep neural networks for forecasting energy demand gives accurate result. Khan A. et al. [30] used machine learning algorithm along with cuckoo search method for forecasting energy requirement. Almalaq, A. et al. [31] used Long short-term memory networks with Genetic Algorithm (GA) to create prediction and optimisation models of energy consumption of buildings. Wen L. et al. [32] used LSTM with particle swarm algorithm (PSO) to correlate load dispatch in a community micro grid with solar power assistance. Similar work is done by Ceylan H. et al. [33] by using GA to estimate energy demand of Turkey using economic indicators.

Some other related works include [34] [35] in which researchers have used PSO to optimally configure the weights of NN to create an accurate model of energy consumption.

Although the reported literature includes the use of GA and PSO, there are several such heuristic algorithms namely, Tabu Search, Simulated Annealing, Travelling Salesman, etc. A review of the algorithms is presented in [36]. Of all these algorithms, GA and PSO are predominantly used because of their exceptional performance when applied to engineering problems. These two methods are broadly similar however they differ in their basic nature of search technique. GA is based on evolution whereas PSO is based on swarm intelligence. A detailed comparison study of these two techniques is presented in [37]. Also, R. Kshirsagar et al. [38] proved that PSO can be derived as special case of GA for a class of engineering problems. GA is a suitable algorithms for non-linear problems.

In this work we used simulation model results of a case study in glass industry to create regression model and an optimization frame work is created using the regression model. The frame work is a multi-objective and multi-constraint based prediction model which is capable of generating input parameter values for a particular output required.

III. REGRESSION MODEL OF HEAT TREATMENT

The case study we considered to analyse is a heat treatment process of glass bottles made of soda-lime material. Cooling part of the heat treatment process is selected for simulation. In the cooling part of the heat treatment we aim to study the changes in material quality and energy consumption in relation to the parameters annealing temperature (^{0}C), Cooling rate (^{0}C/min) and Exit temperature (^{0}C). All the other possible parameters are kept constant for this study. These three parameters are the independent parameters of the process.

For the purpose of creating data related to heat treatment of glass, we have setup maximum and minimum values of independent variables of heat treatment process of glass bottles. The details of the independent parameter values are presented in Table I.

TABLE I. LIST OF INDEPENDENT PARAMETERS AND THEIR VALUES OF HEAT TREATMENT OF GLASS

S.No	Parameters	Level 1	Level 2	Level 3
1	Annealing/Initial Temperature (^{0}C)	545	565	605
2	Cooling rate above S.T (^{0}C/min)	3	6	9
3	Exit Temperature (^{0}C)	70	110	150

Using the values of each independent/input parameters, we have created a full factorial design of experiment which results in 27 experiments. These experiments are carried out in a computer simulation using ANSYS. The simulation is a cooling simulation where the work piece is at annealing temperature and cools down to exit temperature at the given cooling rate. In each simulation, the value of maximum stress and energy required are evaluated. Results of simulation along with input parameters of each simulation are presented in Table II.

TABLE II. SIMULATIONS RESULTS. VALUES OF INPUT AND OUT PARAMETERS

Input parameters			Output parameters (Simulation results)	
Initial temperature (^{0}C)	Cooling rate(^{0}C/min)	Exit temperature(^{0}C)	Max Stress (von-Mises) Pa	Energy (J)
605	6	150	476420	73309
545	6	70	476350	76573
545	9	150	714670	63029
545	3	110	238230	70496
605	6	70	476350	86364
545	3	70	238200	77017
545	3	150	238270	63975
545	6	110	476390	70045
545	6	150	476420	63516
545	9	70	714440	76090
545	9	110	714540	69560
565	3	70	238200	80276
565	3	110	238230	73756
565	3	150	238270	67236
565	6	70	476350	79837
565	6	110	476390	73309
565	6	150	476420	66780
565	9	70	714510	79355
565	9	110	714620	72825
565	9	150	714490	66295
605	3	70	238200	86795
605	3	110	238230	80276
605	3	150	238270	73756
605	6	110	476390	79837
605	9	70	714440	85886
605	9	110	714540	79355
605	9	150	714660	72824

Using these results (listed in Table II), a data set is created with three input parameters namely Initial temperature, Cooling rate and Exit temperature, two output parameters namely maximum stress value and energy. NN is used to create a regression model. Details of the NN is presented in Figure 1.

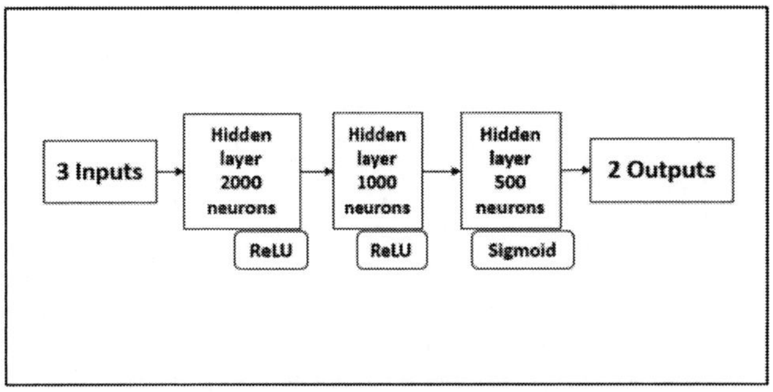

Figure 1. Flow chart of optimisation algorithm.

It can be noted that a specific architecture (as shown in Figure 1) is used in this work to create NN. As explained in [25], there is no logical way to decide the number of neurons and number of hidden layers in NN architecture for a particular type of problem or data set. In this report the architecture shown in Figure 1 is realised after several attempts made, by changing the number of hidden layer, number of neurons in the hidden layers and examining the results after several training iterations. The NN code is set to iterate till the loss function evaluated on training set is same as testing set and percentage error evaluated between predicted values and the actual values is less than 1%. This exercise confirms that the created model is accurate and does not over fit the data.

Data is normalised before feeding it to NN. It is normalised using the formula given by (1).

$$X_{norm} = \frac{X - X_{min}}{X_{max} - X_{min}} \quad (1)$$

Here, X_{norm} – normalised value, X_{min} – minimum value of a particular variable , X_{max} – maximum value of a particular variable , X – actual value of the variable (from Table II). It can be noted that the data in normalised form is unit less.

TABLE III. ACTUAL OUTPUT PARAMETERS, ESTIMATED OUTPUT PARAMETERS AND PERCENTAGE ERROR

Output parameters (Simulation results)		Output parameters (Neural network Model predictions)		Percentage error	
Max Stress (von-Mises) Pa	Energy (J)	Max Stress (von-Mises) Pa	Energy (J)	Max Stress (von-Mises)	Energy
476420	73309	474945	73260.8	0.31%	0.07%
476350	76573	473229	76579.8	0.66%	0.01%
714670	63029	716380	62957.3	0.24%	0.11%
238230	70496	239977	70599.4	0.73%	0.15%
476350	86364	473776	86353.2	0.54%	0.01%

The NN is trained by dividing the data into training and testing set. The division is carried out randomly. 80% of the data is used to create the NN model. Rest of the 20% of the data is used to validate the model created by training data. Mean squared error is measured on both testing and training data set. Training of the NN is carried out till mean squared error measured on both testing and training are equal so that the prediction model does not over fit the data. Percentage error between the predicted values and the actual values is within 1%. Percentage error evaluated over five data points is presented in Table III.

IV. OPTIMISATION MODEL USING HEAT TREATMENT MODEL

Regression model created (in Section III) to predict Von-Mises stresses and energy is used in GA to create optimization frame work. It enables us to find out the values of input parameter values for the required output. In the frame work, objective function is created using regression models created using NN. The objective function is given by (2).

Objective function = Minimize {abs(Stress evaluated by regression model-Required stress value)+abs(Energy evaluated by regression model-Required energy value})

(2)

Note that the parameters used in (2) are in normalized form, hence the issue of incomparable engineering units in a single equation does not arise. Also the parameters used in (2) namely "Required stress" value and "Required energy" value are set by the user. The range of parameters is 0 to 1. The regression model and the optimization process works well for interpolation.

The objective function is created as a minimization problem in which GA tries to find out a stress and energy value which is equal to required stress and energy values. The iterative steps in GA are as follows

- Step 1: Create an initial population. Here the initial population is nx3 matrix where n is the population length and 3 is the number of input variables in GTS data. Each value in the matrix is chosen randomly that lies between 0 and 1. This is because, regression model created is based on normalized data.
- Step 2: Evaluate stress, energy values for each population set (each row of nx3 matrix) using regression model and then evaluate objective function given by (2).
- Step 3: As the objective is to minimize (2), the rows of population set is sorted in ascending order with respect to objective function.
- Step 4: Top half of the population is selected as fit population. Cross over is carried out on the fit population by selecting two random chromosomes. Another half of the population is replaced with newly formed chromosomes.
- Step 5: Mutation is carried out by replacing randomly selected gene in a randomly selected chromosome.
- Step 6: repeat from step 1 to 5 with new set of population formed in Step 5 until the percentage error between the actual vs predicted values is within the set value.

Flow chart of the Algorithm explained in the above 6 steps is presented in Figure 2.

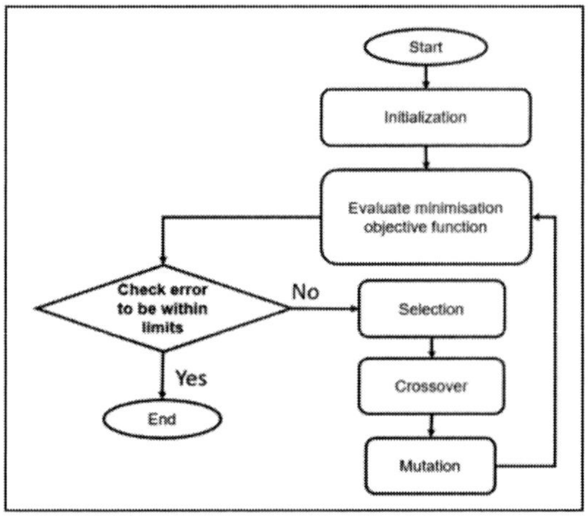

Figure 2. Flow chart of optimisation algorithm.

To start the algorithm, we have set required stress value and energy value as presented on Table IV.

TABLE IV. RESULTS OF OPTIMISATION FRAMEWORK

S. No.	Max Stress (von-Mises) Pa	Energy (J) (Set value)	Temp (GA result)	Exit temp (GA result)	Cooling rate(°C/min) (GA result)	Max Stress (von-Mises) Pa	Energy (J) (NN result)

	(Set value)					(NN result)	
1	333494	74912	593.216	130.294	4.31758	332858	74903.1
2	285847	74912	573.789	112.013	3.74986	286081	74812.2
3	285847	65405.6	548.204	143.704	3.80564	285242	65338.5
4	476435	65405.6	553.14	147.554	6.00404	476384	65379.8
5	667023	84418.4	601.384	75.5933	8.3983	666715	84426.1
6	667023	79665.2	598.011	101.057	8.36115	662141	79657.9

Using trial and error, we identified that a population size of 50 results on definite convergence. So the population size is set to 50 and the algorithm is iterated till the required accuracy is achieved. At the end of the iterations, top chromosome set after sorting is identified to be the best solution. This exercise is carried out for all the 6 set of stress and energy values and corresponding input values obtained using GA are presented in Table IV with label "GA result".

To validate the obtained values using GA, "GA result" set are input to neural network model and stress and energy values are calculated. The values are presented in Table IV with label "NN result".

TABLE V. PERCENTAGE ERROR OF ACTUAL VALUES AND THE VALUES OBTAINED BY OPTIMISATION FRAMEWORK

Optimization frameworks results		Set values		Percentage error	
Max Stress (von-Mises) Pa	Energy (J)	Max Stress (von-Mises) Pa	Energy (J)	Max Stress (von-Mises)	Energy
332858	74903.1	333494	74912	0.19%	0.01%
286081	74812.2	285847	74912	0.08%	0.13%
285242	65338.5	285847	65405.6	0.21%	0.10%
476384	65379.8	476435	65405.6	0.01%	0.04%
666715	84426.1	667023	84418.4	0.05%	0.01%
662141	79657.9	667023	79665.2	0.73%	0.01%

Percentage error of between "GA result" and "NN result" is evaluated and tabulated in Table V. Maximum value of percentage error falls below 1% in this case also.

V. CONCLUSIONS

Heat treatment process of glass is studied in this work. A specific pattern of heat treatment and cooling is required to achieve desired properties in the glass material. For the purpose of study, independent/input parameters list is created with their maximum and minimum values. A full factorial design of experiments set is listed for computer simulations. Output parameters namely stress, energy are evaluated using the simulations. A multi-objective and multi-criteria optimization framework is created using regression model and genetic algorithm. Results obtained using both regression model and optimization model are well within 1% error. Although only one case study is used to carry out the analysis, the optimization framework we proposed can be used for any industry problem.

The framework proposed in this manuscript allows to set the output parameters and evaluate input parameters. GA is used for the evaluation. The framework also allows to constrain the range/value of few input parameters and evaluate others. In this way it is possible to obtain a range of input parameters which fit the operating conditions. So for a particular value of energy consumption and stress value and exit temperature, it is possible to obtain various values of cooling rate and initial temperature. An example is presented in Table VI. In this way, an optimal input parameter-set can be evaluated for a particular operating conditions. Several such input parameter-set can be evaluated using the proposed framework.

TABLE VI. MULTIPLE INPUR PARAMETER-SET FOR A PARTICULAR VALUES OF OUTPUT PARAMETERSSS

Initial temperature (°C)	Cooling rate(°C/min)	Exit temperature(°C)	Max Stress (von-Mises) Pa	Energy (J)
545.971	4.80058	112.333	381141	70158.8
547.31	4.79042	113.755		
552.531	4.86698	119.288		
566.68	4.76859	132.511		
574.075	4.88605	139.3		

ACKNOWLEGMENTS

Work in this manuscript is part of the Innovate UK funded project entitled "AI lean six-sigma process optimization for energy efficiency and waste reduction in foundation industries". We thank Innovate UK for their support.

REFERENCES

[1] K. Rajkumar and S.Aravindan "Tribological performance of microwave sinteredcopper–TiC graphite hybrid composites", Tribology International 44 (2011) 347–358.

[2] J. William and R. Mehl, "Reaction kinetics in process of nucleation and growth", Trans. Metall. Soc, AIME, vol. 135, pp. 416-442, 1939, doi:10.1007/s11661-011-0780-2.

[3] M. Avrami, "Granulation, phase change, and microstructure kinetics of phase change. Iii," The Journal of chemical physics, vol. 9, no. 2, pp 177-184, 1941, doi: 10.1063/1.1750872.

[4] M. Todinov, "On some limitations of the Johnson-mehl-avrami-kolmogorov equation", Acta Materialla, vol. 48, no. 17, pp. 4217-4224, 2000, doi:10.1016/S1359-6454(00)00280-9.

[5] M. Maisuradze, Y. V. Yudin, M. Ryzhkov, "Numerical simulation of pearlitic transformation in steel 45kh5mf", Metal Science and Heat Treatment, vol. 56, no. 9-10, pp. 512-516, 2015, doi: 10.1007/s11041-015-9791-8.

[6] K. Sadeghipour, J. Dopkin, K. Lí, "A computer aided finite element/experimental analysis of induction heating process of steel," Computers in industry, vol. 28, no. 3, pp. 195–205, 1996, doi:10.1016/ 0166-3615(95)00072-0.

[7] Z. Guo, N. Saunders, J. Schillé, and A. Miodownik, "Material properties for process simulation," Materials Science and Engineering: A, vol. 499, no. 1-2, pp. 7-13, 2009, doi:10.1016/j.msea.2007.09.097.

[8] T. Reti, Z. Fried, I. Felde, "Computer simulation of steel quenching process using a multi-phase transformation

model," Computational Materials Science, vol. 22, no. 3-4, pp. 261-278, 2001, doi:10.1016/ S0927-0256(01)00240-3.

[9] G. Li, H. X. Lu, X. G. Hu, Q. Zhu, "Numerical simulation of slurry making process of 7075 aluminum alloy under electromagnetic field in rheocasting process," in Solid State Phenomena, vol. 285. Trans Tech Publ, pp. 373–379, 2019. doi:10.4028/www.scientific.net/SSP.285.373.

[10] Ramprasad R, Batra R, Pilania G, et al. "Machine learning in materials informatics: Recent applications and prospects". npj Comput Mater, no. 3: pp. 54, 2017

[11] Xue D, Xue D, Yuan R, et al. "An informatics approach to transformation temperatures of NiTi-based shape memory alloys". Acta Mater, no. 125, pp. 532–541, 2017.

[12] Huber L, Hadian R, Grabowski B, et al. "A machine learning approach to model solute grain boundary segregation". npj Comput Mater, no. 4: pp. 64, 2018.

[13] Zhu Q, Samanta A, Li B, et al. "Predicting phase behavior of grain boundaries with evolutionary search and machine learning". Nat Commun, no. 9, pp. 467, 2018.

[14] Raccuglia P, Elbert K C, Adler P D F, et al. "sMachine-learning-assisted materials discovery using failed experiments". Nature, no. 533, pp. 73–76, 2016.

[15] Agrawal A, Deshpande P D, Cecen A, et al. "Exploration of data science techniques to predict fatigue strength of steel from composition and processing parameters". Integrating Mater, no. 3, pp. 90–108, 2014.

[16] Agrawal A, Choudhary A. "An online tool for predicting fatigue strength of steel alloys based on ensemble data mining". Int J Fatigue, no. 113, pp. 389–400, 2018.

[17] Xiong, Jie, TongYi Zhang, SanQiang Shi. "Machine learning of mechanical properties of steels." Science China Technological Sciences no. 63.7, pp. 1247-1255, 2020.

[18] Masai, Hirokazu, et al. "Examination of structure and optical properties of Ce3+-doped strontium borate glass by regression analysis." Scientific reports, no. 11.1 , pp. 1-12, 2021.

[19] Samuel, Bassey Okon, Malachy Sumaila, Bashar Dan-Asabe. "Manufacturing of a natural fiber/glass fiber hybrid reinforced polymer composite (PxGyEz) for high flexural strength: An optimization approach." The International Journal of Advanced Manufacturing Technology, no. 119.3, pp. 2077-2088, 2022.

[20] Sathisha N, Hiremath SS, Shivakumar J "Prediction of material removal rate using regression analysis and artificial neural network of ECDM process". Int J Mech, no. 3(2), pp. 69–81, 2014.

[21] Patel P, Sheth S, Patel T "Experimental analysis and ANN modelling of HAZ in laser cutting of glass fibre reinforced plastic composites". Proc 3rd Inter Conf Innov Aut Mechatronics Eng, no. 23, pp. 406–413, 2016.

[22] Nasser ESA, Elkaseer A, Nassef A "Abrasive jet machining of glass: experimental investigation with artificial neural network modelling and genetic algorithm optimization". J Cogent Eng, no. 3(1): pp. 1276513, 2016.

[23] Bezerra EM, Ancelotti AC, Pardini LC, Rocco JAFF, Iha K, Ribeiro CHC "Artificial neural networks applied to epoxy composites reinforced with carbon and E-glass fibers: analysis of the shear mechanical properties." Mater Sci Eng, no. 464, pp. 177–185, 2017.

[24] Bebis, George, Michael Georgiopoulos. "Feed-forward neural networks." IEEE Potentials, no. 13.4, pp. 27-31, 1994.

[25] D. Chen, D. Sun, J. Fu, S. Liu, ''Semi-supervised learning framework for aluminum alloy metallographic image segmentation,'' IEEE Access, vol. 9, pp. 30858–30867, 2021.

[26] A. Karami, G. H. Roshani, E. Nazemi, S. Roshani, ''Enhancing the performance of a dual-energy gamma ray based three-phase flow meter with the help of grey wolf optimization algorithm,'' Flow Meas. Instrum., vol. 64, pp. 164–172, Dec. 2018.

[27] G. H. Roshani, R. Hanus, A. Khazaei, M. Zych, E. Nazemi, V. Mosorov, ''Density and velocity determination for single-phase flow based on radiotracer technique and neural networks,'' Flow Meas. Instrum., vol. 61, pp. 9–14, Jun. 2018.

[28] Cai, M.; Pipattanasomporn, M.; Rahman, S. "Day-ahead building-level load forecasts using deep learning vs. traditional time-series techniques". Appl. Energy, no. 236, pp. 1078–1088, 2019.

[29] Chitalia, G.; Pipattanasomporn, M.; Garg, V.; Rahman, S. Robust short-term electrical load forecasting framework for commercial buildings using deep recurrent neural networks. Appl. Energy, no. 278, pp. 115410, 2020.

[30] Khan, A.; Chiroma, H.; Imran, M.; Bangash, J.I.; Asim, M.; Hamza, M.F.; Aljuaid, H. "Forecasting electricity consumption based on machine learning to improve performance: A case study for the organization of petroleum exporting countries (OPEC)". Comput. Electr. Eng, no. 86, pp. 106737, 2020.

[31] Almalaq and A.; Zhang, J.J. "Evolutionary deep learning-based energy consumption prediction for buildings." IEEE Access, no. 7, pp. 1520–1531, 2018.

[32] Wen, L.; Zhou, K.; Yang, S.; Lu, X. Optimal load dispatch of community microgrid with deep learning based solar power and load forecasting. Energy, no. 171, pp. 1053–1065, 2019.

[33] H. Ceylan, H. Ozturk, Estimating energy demand of Turkey based on economic indicators using genetic algorithm approach, Energy Conversion and Management no. 45, pp. 2525–2537, 2004.

[34] J. Kennedy and R. Eberhart, ''Particle swarm optimization,'' in Proc. Int. Conf. Neural Netw. (ICNN), vol. 4, pp. 1942–1948, 1995.

[35] H. R. Madvar, M. Dehghani, R. Memarzadeh, E. Salwana, A. Mosavi, S. Shahab, ''Derivation of optimized equations for estimation of dispersion coefficient in natural streams using hybridized ANN with PSO and CSO algorithms,'' IEEE Access, vol. 8, pp. 156582–156599, 2020.

An Islanded Community Solar Microgrid with Capability of Future Fractal Growth

Mark Apperley
Department of Software Engineering
University of Waikato
Hamilton, Aotearoa New Zealand
email: mark.apperley@waikato.ac.nz

Tama Toki
Aotea Energy
Auckland, Aotearoa New Zealand
email: tama@aoteaenergy.com

Abstract— **A design for a non-grid-connected (islanded) smart community microgrid is developed and elaborated. This focuses on a real community, and the design is developed in such a way as to take into account current energy demands, and future expansion (given that currently there is no formal electricity supply to the community), but also, since the community is located on an island with a number of neighbouring communities and industries, the design is developed to enable future fractal growth of the micro-grid. To meet these requirements, the development covers needs analysis, the micro-grid configuration, and the initial sizing of the various components. Configuration and sizing are then optimised for the initial microgrid, taking into account the particular community social infrastructure characteristics and dynamics.**

Keywords-community microgrid; fractal microgrid design.

I. Introduction

This paper describes the development of a design for a smart electricity microgrid in Aotearoa New Zealand, for the community of Motairehe on Aotea/Great Barrier Island, a remote island approximately 100km northeast of Auckland. There is no reticulated power system on Aotea. The entire population of the island (~1000) live off-the-grid, running their own solar/battery power systems, which are supplemented by petrol or diesel powered generators, natural gas and wood fires, and in almost all cases, the solar/battery systems do not provide nearly enough of the households' energy needs, so there is a heavy reliance on the fossil-fuel powered back-up generators.

This absence of existing infrastructure provides an opportunity to improve the lives of the island's population, and contribute to New Zealand's efforts to reduce carbon emissions and expand clean energy use [1]. The community on which this design focuses contains the majority of the indigenous Māori population of Aotea, approximately forty households and 90-100 people. The initial design is for the central part of the community, ten households together with a *marae*, but it is intended that through a phased approach, the microgrid would be extended over time to include the remainder of the households, and then on into neighbouring communities. A *marae* is a place where traditional *Māori* ceremonies and meetings are held. It normally comprises a meeting space, *marae ātea*, a meeting house, *wharenui*, and a kitchen/dining room, *wharekai*. The *marae* can accommodate *hui*, either short, smaller meetings, or larger *iwi* (tribal) events that may run for several days, with accommodation being provided in the *wharenui*.

To accommodate this anticipated expansion, focussing on the use of non-dispatchable and distributed renewable energy generation (solar panels), and to some extent exploiting the absence of any existing legacy grid, the design described is based on a fractal grid model [2]. Considerations in this design include distributing both generation and storage across the community, in such a way as to ensure maximum local consumption of locally produced energy, to enable and support community utilisation of all energy produced, yet to minimise the required inter-node transmission capacity. A further aspect is to ensure community engagement, and essentially "ownership" of the microgrid [3].

While there is a growing body of literature on the design of community microgrids e.g., [4][5][6], the work described here essentially begins from first principles in order (i) to recognise the islanded environment, (ii) to base the design principally on solar power and batteries, and (iii) to acknowledge the social context of the development.

The paper is structured in the following way. Section II develops appropriate household and *marae* load profiles, necessary because there is no existing reticulated power system. These load profiles are then used in Section III to establish what would be an adequate or appropriately sized solar/battery configuration for a household, and for the *marae*. This configuration is then adjusted in the context of a fractal microgrid design in Section IV, which clearly demonstrates the advantages of this concept. Section V provides a discussion of the optimal configuration and distribution of resources in the microgrid. The overall design, and the current state of the development, is then discussed in the concluding Section VI.

II. Load Profile Estimation

Annual load profiles were established for ten individual households. As there is currently no grid supply to the community, then these profiles were derived from real households with appropriate occupancy and appliance utilisation characteristics, located at a similar latitude on the New Zealand mainland. Those adopted were selected on the basis that the houses did not have electric hot-water systems, nor did they use electricity as their primary form of heating. These were also households comprising a range of occupant numbers, both adults and children. The consumption profiles for these houses varied considerably, in terms of average daily use, hourly use over the day, and seasonal differences. The data used was hourly data over a whole year – 8760 data points for each house [7].

While the variation between households was retained, the profiles were scaled to give an average daily base load (across the ten houses) of 7.5 kWh. This data is summarised in Table 1.

TABLE I. SUMMARY OF HOUSEHOLD BASE LOAD DATA

ID	*#Adults*	*#Child*	*Daily average*	*Scaled Average*	*Comments*
A1	2	0	11,714	7,994	Gas heating
A2	2	0	10,981	7,493	Fossil heating
A3	2	0	8,956	6,112	Gas heating
A4	2	0	14,047	9,586	Gas heating
A5	1	0	5,840	3,985	Gas heating
A6	4	1	13,509	9,219	Gas heating
A7	2	0	11,150	7,609	Gas heating
A8	3	1	10,249	6,994	Gas heating
A9	2	0	6,990	4,770	Gas heating
A10	2	0	16,461	11,233	Gas heating
Average daily consumption			**10,990**	**7,500**	

A load profile for the *marae* was also established using similar techniques, based on the occupancy level. One of the household profiles (A4, Table 1) was used as the starting point, but the scaled data was doubled to represent 4 adults (estimated permanent *marae* occupancy), meaning the base load daily average is 19.171 kWh. To this base profile was then added a randomly generated *hui* load. Fourteen *hui* were added over the year, with their start dates randomly generated, and with randomly allocated durations of 1 to 3 days. This was based on suggestions from the community of typical *hui* frequency and size. An estimated supplementary load profile was created for a *hui*, which included additional cooking, lighting, and heating during the event. This added load runs from 2.00pm to 2.00pm, so assumes that even a 1-day *hui* involves an overnight stay for the non-local participants.

For the *marae*, the addition of these *hui* increases the annual average daily consumption to 24.275 kWh, and the maximum consumption in any one hour over the year from 6.051 to 9.465 kW.

III. INDIVIDUAL HOUSEHOLD DESIGN

Based on these load profiles, and standard design guidelines for solar installations in New Zealand:

- solar panel capacity ~ average daily load / 4
- battery capacity ~ average daily base load x 3

an initial analysis was carried out for the ten households and the *marae*.

For this initial analysis, with each household, and the *marae*, operating as an independent unit, the standard configuration used was solar panels of 4 kW capacity, plus a battery of 22.5 kWh capacity, for each site. Annual solar data for Aotea was used to calculate the solar panel output for each hour of the year [8], and key points from this analysis of the non-networked configuration are shown in Table 2.

A discretionary or divertible load was added in for this analysis. In reality, this might, for example, represent hot-water heating or EV charging, electricity usage carried out when there is a surplus, over and above the base load. This discretionary load was set to a daily maximum (per household) of 5 kWh, and was invoked only when the battery was at 95% charge level or more, and there was surplus solar generation. This produced an overall total average discretionary load across the community of 30.4 kWh per day, and there were just 9 days in which no discretionary load was possible at all.

TABLE II. SUMMARY OF INITIAL NON-NETWORKED SIMULATION

ID	*Average Daily Base Demand (W)*	*Peak Daily Base Demand (W)*	*Average Daily Discretionary Load (W)*	*Total Surplus Hours*	*Total Failed Hours*	*Average Daily Failed Supply (W)*
A1	7,994	16,406	3589	1604	16	13
A2	7,493	15,281	3779	1704	12	14
A3	6,112	9,087	4016	1939	0	0
A4	9,586	27,908	3093	1365	143	217
A5	3,985	11,730	4364	2200	0	0
A6	9,219	17,160	3364	1374	127	142
A7	7,609	19,738	3735	1663	58	47
A8	6,994	14,018	3778	1750	0	0
A9	4,770	11,893	4229	2057	0	0
A10	11,233	31,066	2899	1129	317	591
Marae	24,275	106,162	808	274	2184	8908

It is interesting to note (Table 2) that under this autonomous mode of operation, and with this configuration, for four of the households there were no base load supply failures at all over the entire year. For the other six households, there were times when their systems were unable to meet the basic household needs, with House 10 showing the most extreme case of 317 hours of failed base load supply. The *marae*, because of its very high peak demands during *hui*, showed 2184 hours in which the system was unable to meet its demands, and with this shortfall averaged over the whole year, that amounts to nearly 9kWh per day. By contrast, all sites, including the *marae*, showed a significant number of hours in which there was an unused surplus of solar energy.

Of course the failure occurrences for the individual houses do not all occur together, although after a day of low sunshine, it is more likely that such events may coincide. For all of those failed hours across the ten households shown in Table 2, these were spread across 355 hours of the year.

To give an idea of the energy balance situation, in Figures 1 and 2 the annual hourly balance is shown for two extreme houses, House 5 (zero failed hours) and House 10 (317 failed hours). For these energy balance plots, the consumed energy includes battery charging as well as the actual domestic loads, and the produced energy comprises both solar generation and anything delivered from the batteries [9]. Perfect balance is shown by points on the diagonal. For House 5 (Figure 1), perfect balance is achieved 75% of the time. The very evident stack of points close to the vertical axis represent hours of sunlight (high generation) at times of low load, when the batteries are fully charged. The other vertical cluster of points at a load of 2.2kW represent those daylight hours when the discretionary load has been activated, but there is still some surplus. House 10 (Figure 2) actually achieves perfect balance 83% of the time, but of course, as can be seen from the graph, there are 317 hours (3.6%) when the full demand is not met.

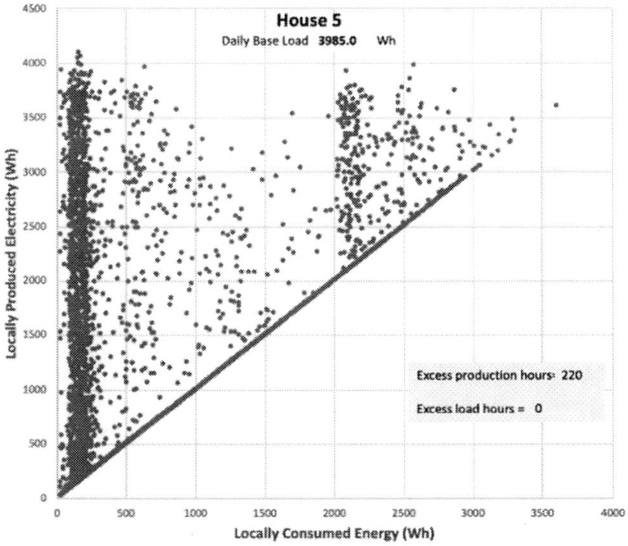

Figure 1. Hourly energy balance over the full year for House 5 in the non-networked mode.

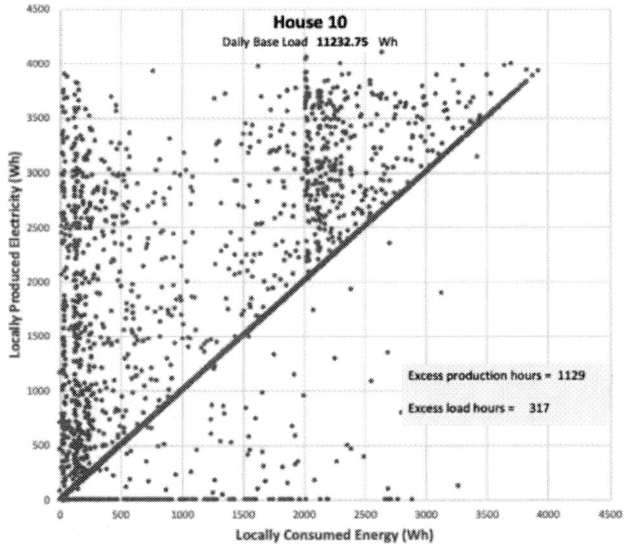

Figure 2. Hourly energy balance over the full year for House 10 in the non-networked mode.

These two houses represent the extremes of generation/load balance using this basic design; all of the others lie between these in their energy balance characteristics [9].

IV. Fractal Microgrid Design

Clearly with demands and surpluses varying between households, and the occasional peak demand from the *Marae* during *hui*, community sharing has potential to achieve a greater degree of balance between generation and demand. A fractal microgrid design [2] as represented in Figure 3 will now be considered.

In this model, any terminal node (A to F, and I in Figure 3) can comprise any combination of load, generation, and/or storage. Typically, these would represent individual households such as those being considered here, but they could also represent a community solar panel array (generation only), a community service, such as street lighting or EV charging (load only), or a community battery (storage only). Some of the households might not include solar panels and/or storage. The non-terminal nodes (G, H, and J in Figure 3) will always present themselves as intelligent and active to the higher level grid they are connected to, and will comprise a combination of load, generation and storage, but from the perspective of the higher level grid, will appear as a single entity with these properties, hence the term, *fractal* microgrid [10].

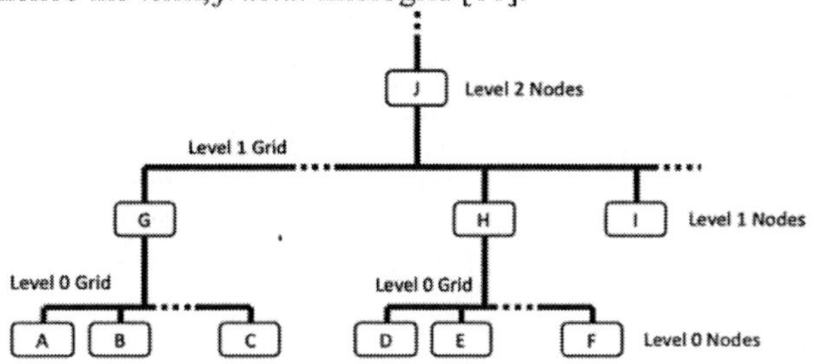

Figure 3. The fractal microgrid concept.

The fractal microgrid proposed here initially comprises just a single level 0 grid (Figure 3) and no higher levels. The *marae* forms the level 1 node (node G in Figure 3), and the community households the level 0 nodes (nodes A, B, C in Figure 3). A more specific representation of this Motairehe fractal microgrid is shown in Figure 4.

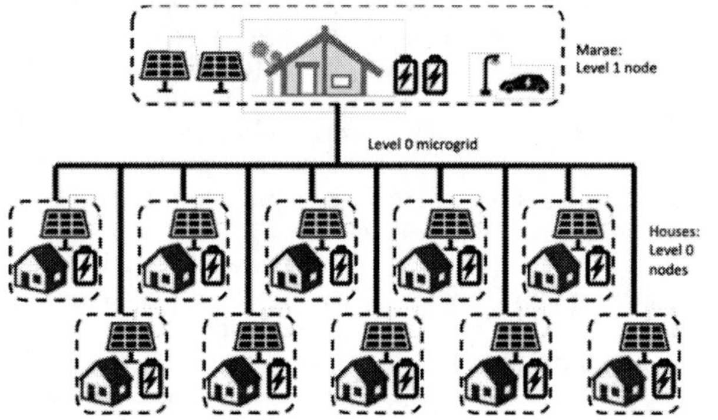

Figure 4. The proposed Motairehe fractal microgrid comprising ten houses and the *marae*.

In this model, it can seen that each household retains local solar generation and battery storage, but is also connected to the microgrid. This means that at times of surplus, the household can contribute to the microgrid, potentially supporting neighbours, and/or the *marae*, and at times of shortfall, it may be able to draw from neighbours or the *marae* surplus. The *marae* similarly is connected to the microgrid, and retains its local solar generation and storage. Also shown in the *marae* set up is the notion of community services – in this case, street lighting and EV charging. Such services could also be simply connected to the microgrid itself (rather than be part of the *marae* node), as could, for example, additional community solar panels.

Although shown here as a single community system, and not explicitly demonstrating a fractal structure itself, it has been developed in this way consistent with the fractal microgrid concept, in order that it could in the future be readily extended to:

- multiple community grids within the same general area, each connected to the *marae* as the upper-level node; or
- the *marae* itself could be connected to a higher level node, potentially bringing together more distant groupings.

Possible future extensions to the microgrid of this nature would follow the general fractal model of Figure 3.

The approach taken with the initial design of the microgrid is to consider overall the same total solar generation and the same total battery storage as was used for the non-networked analysis of Sections II and III. At this stage, it is assumed:

- All houses will have the same sized system of solar panels and batteries;
- The *marae* may have a different configuration, and the design will attempt achieve a distribution which minimizes the overall energy transfer between the *marae* and the community, in either direction;
- It is anticipated that the fractal microgrid should achieve an overall better utilisation of the generated electricity, reducing over the entire site both the wasted excess production and the failed supply.

Initially, it is a simple matter of considering the total generation capacity, the total load, and the total storage, across the whole community and *marae*, and carrying out an hourly energy balance analysis for the whole year [9]. However, once that analysis has been carried out, then the distribution of generation and storage between the houses and the *marae* needs to be explored to minimize the overall grid flow, since a higher grid flow will require a more substantial cable, and/or imply greater transmission losses.

The results of the initial analysis are shown in the energy balance plot of Figure 5. The impact of *hui*, which overall present a significant load, can be readily seen, as the hours which correspond to *hui* at the *marae* are highlighted in this plot. What is remarkable, is that by considering the total generation, the total load, and the total battery capacity, the overall number of shortfall hours over the year for the whole community have been reduced to 64, significantly less than the 2184 previously experienced by the *marae*, and the 355 by all of the houses together. In fact, only 16 of the shortfall hours actually coincide with *hui*.

Of course, this plot (Figure 5) does represent just the base load; any discretionary load would need to be taken during those hours of surplus, above the diagonal. Referring back to Table 2, it can be seen that these 1870 hours of overall surplus are more than those experienced by some of the houses in the non-networked mode. The impact of the fractal network on discretionary load is discussed later, in relation to Figure 8. However, each site (both Level 0 and Level 1) would need "smart" control to prioritise demand in the following sequence:

(i) Local base load;
(ii) Local battery charging;
(iii) Microgrid (community) needs;
(iv) Household discretionary load.

To determine (iii) and (iv) will require more sophisticated software at each site, and communication between sites, than is conventionally seen in isolated off-grid solar sites, which do not have to concern themselves with (iii).

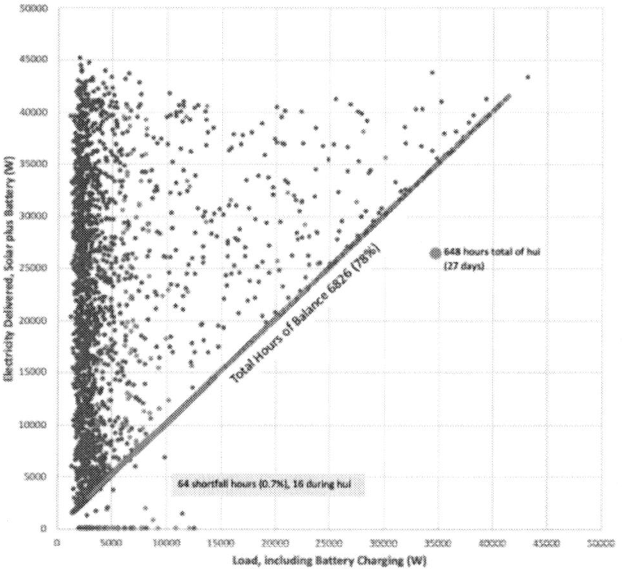

Figure 5. The overall energy balance for the fractal microgrid model.

It is clear from this initial analysis of the fractal community microgrid model, that overall a much better utilisation of the generated electricity has been achieved, reducing, across the entire site, both the excess production and the failed supply. This has utilised the same overall total solar panels, and the same overall total battery storage, as with the original non-networked model of Section III. These numbers are:

- *Solar panels:* For the non-networked model, each house, and the *marae*, was fitted with ten 400W panels, giving a total of 110 x 400W.
- *Batteries:* For the non-networked model, each site was provided with 22.5 kWh of battery, leading to a total battery capacity of 247.5 kWh.

Before proceeding to optimise the configuration by attempting to minimise grid flow, it is worth considering other aspects of this fractal microgrid model. Figure 6, for example, shows the daily variation in total base load for the community, and highlights those days which correspond to *hui*. While it can be seen that there are daily peaks corresponding to *hui*, there are also peaks which are not associated with *hui*.

In Figure 7, the nine days over the year in which supply shortfall occurs are shown, again with correspondence to *hui* highlighted. Three of these days do coincide with *hui*, but there are three quite significant shortfall days which do not. Those days are in the middle of winter, and in this data set represent days of very low solar radiation.

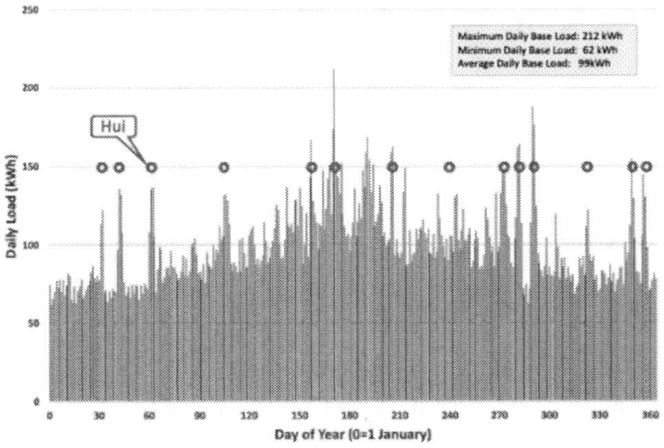

Figure 6. Overall daily base load variation for the whole community, with *hui* days highlighted.

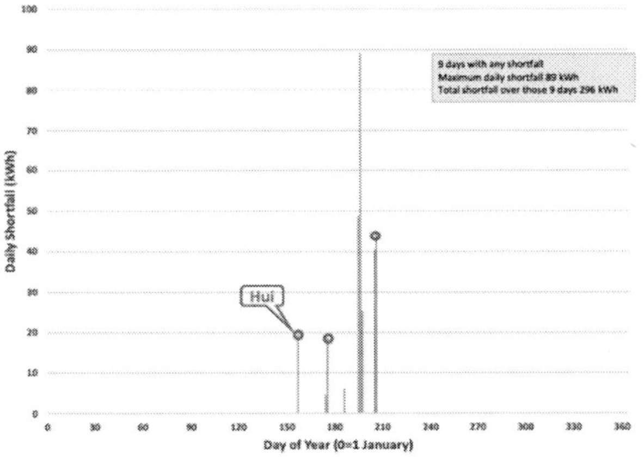

Figure 7. Days of overall shortfall for the community over the year, showing some coincidence with winter *hui*.

Figure 8 shows the daily baseload surplus, which is, of course, highly relevant in relation to discretionary loads. Here, it is evident that the surplus is quite variable, and also that there are periods during the winter where there is no surplus for several days in sequence. For comparison, *hui* are also identified on this plot. For the earlier non-networked configuration, there were just nine days when no discretionary load was available; under the fractal model, there are 60 days. However, the average daily availability of discretionary power is 91 kWh; the allocated average daily load for the non-networked model was 30 kWh. Not surprisingly then, the fractal model offers this option to the whole community on fewer days of the year, because it is making more effective use of surplus generation in providing base load for the whole community, but in total, over the year, there is more than adequate discretionary capability.

It should also be noted that many of the houses, and the *marae*, already have back-up generation resources, and are likely to retain those when the fractal solar microgrid comes into existence. While the household back-up systems are likely to be used only for the individual households, the *marae* back-up would potentially benefit all when needed, covering those days of community shortfall shown in Figure 8.

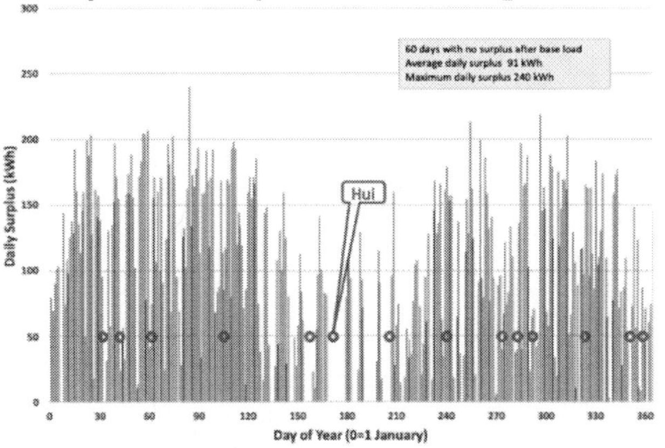

Figure 8. Daily baseload surplus for the community.

This analysis has shown the real adbvantage of the community microgrid approach, in that each houshold is fitted with the same configuration of panels and batteries, and in all cases has a more stable supply of electricity.

V. RESOURCE DISTRIBUTION FOR OPTIMAL MICROGRID OPERATION

Now that the positive aspects of the fractal microgrid approach have been established, it is necessary to consider the optimal configuration of solar panels and batteries. As suggested earlier, it is assumed that all houses have the same configuration, and that the distinction will be between the houses and the *marae*. It is the electricity flow in the microgrid connection which will be the determining aspect of the distribution of resources. For this analysis, it is necessary to explore both:

- the peak flow, since this will dictate the size of cable required, and
- the average flow, since this will determine the overall energy losses in the cable.

The overall resources on which the results shown in Section IV were established, comprise a total of:

- 110 x 400 W solar panels, and
- 247.5 kWh of battery storage.

As an example of the nature of the energy flows between the *marae* and the houses, Figure 9 shows the hourly flow over a year with 20% of the panels and 20% of the battery capacity held at the *marae*. It is evident from this graph that at different times, energy flow can be in either direction, and that the peak flows (largest is ~9.5 kW) occur *to* the *marae*, although overall, more energy appears to flow *from* the *marae*. Almost all of the flows to the *marae* do coincide with *hui*, as can be seen from the graph.

In order to determine the optimum allocation of resources to the *marae*, an analysis was carried out with various proportions of those resources, ranging from 18% to 30%. The results of this analysis, shown in the parallel coordinate plot of Figure 10, reveal clearly that a 20% allocation produces a minimum of both peak grid flow, and daily average grid flow.

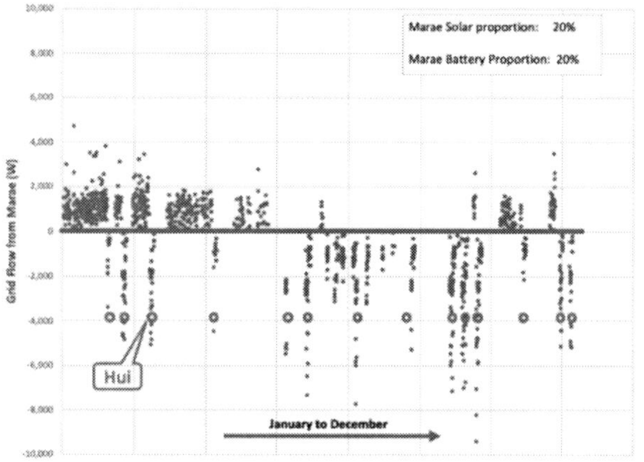

Figure 9. Hourly grid flow from the *marae* to the houses with 20% of generation and storage allocated to the *marae*.

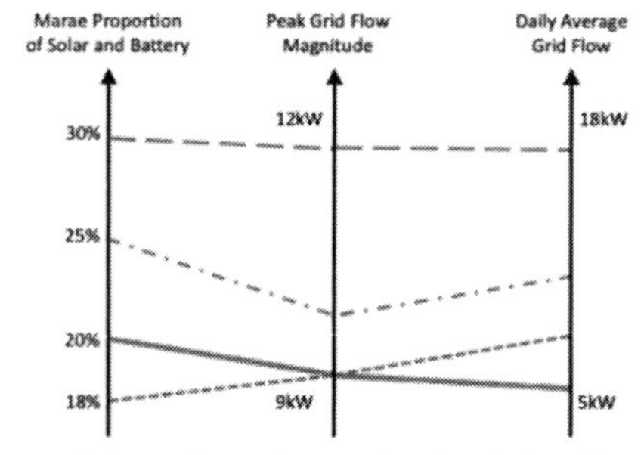

Figure 10. Parallel coordinate plot showing the relationship between the proportion of the resources located at the *marae* and the peak and daily average grid flows.

Finally, to give a better picture of these grid flows, Figure 11 shows (for this 20% *marae* allocation) the distribution of the hourly grid flows over a year.

It can be noted that the grid flows to the *marae* (negative in Figures 9 and 11) occur on far fewer hours than the *marae* showed a shortfall in the non-networked model (see Table 2). But of course, under this optimised fractal microgrid model, the *marae* does have a larger allocation of solar generation and batteries. It should also be noted that this analysis has examined the flow between the *marae* and the houses as a group (refer to Figure 3), not the flow between the houses themselves. These latter flows are likely to be less, and it is reasonable to use the *marae* flow as the basis for cable sizing.

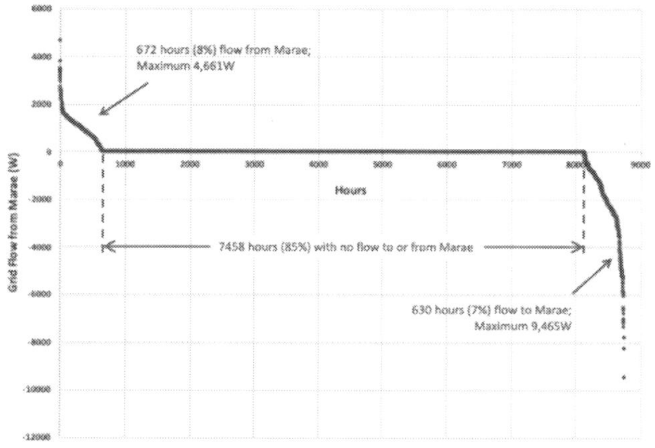

Figure 11. Duration plot of grid flow from the *marae* over a year.

This proposed configuration is elaborated in more detail in the following concluding section.

VI. CONCLUSION AND FUTURE WORK

The analysis of the preceding sections has clearly shown that the suggested fractal solar microgrid model for the Motairehe *marae* and community provides a superior outcome to isolated individual systems, for exactly the same overall solar panel and battery resources. A more resilient and reliable electricity supply for the community ensues, with far fewer hours overall of electricity shortfall.

Based on the numbers from this analysis, the suggested configurations are:

Marae:	8800 W solar panels
	49.5 kWh battery
Houses:	3520 W solar panels
	19.8 kWh battery

These numbers, of course, need to be rounded up for sensible use of currently available technology.

In addition to this basic technology, the houses and the *marae* need to be connected by an appropriate cable capable of carrying at least the calculated 10kW load with minimal losses. A preliminary calculation, taking into account the relatively close proximity of the households and *marae* at Motairehe, suggests that the cost of the interconnecting microgrid cables to handle this 10kW load could have provided ~15% increased solar panel capacity across the site, if no microgrid was included. However the advantages of the microgrid interconnection shown in this analysis, far outweigh the benefits of increased, but isolated, generation. Additional control logic will need to be incorporated within each node of the microgrid, to manage the interaction, particularly in relation to grid needs and offers, and local discretionary demands.

It should also be re-iterated that the adoption of a fractal microgrid model readily enables future expansion, without significant reconfiguration, for example:

- Addition of further community facilities, such as community lighting, shared solar arrays, community EV charging, and community storage, on the level 0 grid.
- Addition of other close-by communities or groups of houses into the system, either as an extension of this level 0 microgrid, or as another level 0 microgrid (refer to Figure 3).
- Expansion of the fractal structure to Level 2 or higher (Figure 3) potentially linking it with more distant communities.

The system as described is currently under development, with support from the MBIE *Māori Housing Renewable Energy Fund* [11].

ACKNOWLEDGEMENTS

The research described in this paper has been supported in part by the MBIE Endeavour Programme, *Public Housing and Urban Regeneration: Maximising Well-being* (Grant ID:20476 UOOX2003), and also the the MBIE Strategic Science Investment Fund grant, *Ahuora: Delivering sustainable industry through smart process heat decarbonisation* (Grant ID: CONT-69342-SSIFAETP-UOW).

REFERENCES

[1] *Emissions reduction targets*. Ministry for the Environment; https://environment.govt.nz/what-government-is-doing/areas-of-work/climate-change/emissions-reduction-targets/ Accessed 2023.02.13.

[2] M. Apperley, "Modelling Fractal-Structured Smart Microgrids: Exploring signals and protocols", *Energy 2019 Conference Proceedings*, IARIA, pp. 13–17. 2019.

[3] A. L. Berka, J.L. MacArthur, and C. Gonelli, "Explaining inclusivity in energy transitions: Local and community energy in Aotearoa New Zealand", *Environmental Innovation and Societal Transitions*, **34**, pp. 165-182, 2020.

[4] L. Che, X. Zhang, M. Shahidehpour, A. Alabdulwahab, and A. Abusorrah, "Optimal interconnection planning of community microgrids with renewable energy sources." *IEEE Transactions on Smart Grid* **8**, pp. 1054-1063, 2015.

[5] M. A. Hossain, H. R. Pota, S. Squartini, F. Zaman, and J. M. Guerrero. Energy scheduling of community microgrid with battery cost using particle swarm optimisation. *Applied Energy*, 254, p. 113723, 2019.

[6] N. Tomin, V. Shakirov, A. Kozlov, D. Sidorov, V. Kurbatsky, C. Rehtanz, and E.E. Lora, (2022), Design and optimal energy management of community microgrids with flexible renewable energy sources, *Renewable Energy*, pp. 903-921, 2022.

[7] M. Apperley, P. Monigatti, and J. Suppers, "Grid-Lite: A network integrated semi-autonomous local area electricity system", In *Proceedings of the 4th International Conference on Green IT Solutions (ICGreen 2015)*, Milan, Italy, SciTePress, pp. 27-33, 2015.

[8] J. Suppers and M. Apperley, "Interactive solar panel simulation tool - from GHI to PV output". In C. Beckmann & T. Gross (Eds.), *INTERACT 2015 Adjunct Proceedings: Fostering Smart Energy Applications Workshop*, pp633–642. Bamberg, Germany: University of Bamberg Press. 2015.

[9] M. Apperley, "Modelling energy balance and storage in the design of smart microgrids", *Energy 2017 Conference Proceedings, IARIA*, Barcelona, pp. 40-45, 2017.

[10] B. B. Mandelbrot, *Fractals: form, chance, and dimension*, W. H. Freeman: San Francisco, CA., 1977.

[11] *Māori and Public Housing Renewable Energy Fund*. https://www.mbie.govt.nz/building-and-energy/energy-and-natural-resources/low-emissions-economy/energy-efficiency-in-new-zealand/maori-and-public-housing-renewable-energy-fund/maori-housing-renewable-energy-fund. Accessed 2023.02.13

The Implementation of the Automatic Dispatching System (ADS) to Support the Smart Grid Pilot Project for Distribution Grid Improvement in Sumba Island

Dimas Bangun Fiddiansyah
Digital Management Division
PT PLN Head Office
Jakarta, Indonesia
email : dimas.bangun@pln.co.id

Dr. Hanny Benchmark
USAID SINAR
Jakarta, Indonesia
email :
hanny.berchmans@tetratech.com

Abstract — **In order to minimize the oil fuel consumption, in its power generation, PLN (Perusahaan Listrik Negara) as an electric state company has promoting a de-dieselization program which either integrated PV and existing diesel into Hybrid or install PV in the isolated system, to get optimum performance and reliability. However, the intermittent characteristic of PV should be considered since this would impact the frequency instability of the network system itself. To maintain the stability of the system, new system operations should be adopted. Thus, ADS (Automatic Dispatching System) is introduced to compensate the fluctuation sources from PV. Before ADS was implemented, dispatch combination between Diesel Generator (DG), and Photovoltaic (PV) has triggered unreliable supply since there is no load control center as well as the insufficient communication channel available. Implementing ADS, this allows PLN to minimize fuel consumption, and develop the first operational Smart Grid system. Subsequently, ADS has proven capable of maintaining frequency stability in the system that contains intermittent power generation. Based on the simulation study, and actual measurement testing, it can maintain the stability frequency within in its normal range (49,5 < Hz < 50,5).**

Keywords — Automatic Dispatching System, Diesel Generations, Photovoltaic, Smart Grid system

I. INTRODUCTION

In recent years, PV based generating systems have been widely implemented into distribution systems due to environmental concerns. VRE (Variable Renewable Energy), such as PV and wind energy may bring additional reliability benefits to the system, as they can be operated along with the DG units, thereby minimizing customer interruptions, in case of system emergencies [9]. However, PV power output is not constant and tends to fluctuate depending on weather conditions. This fluctuating power causes frequency deviations and reduction in reliability of the isolated power utility or microgrid when large output power from several PV systems is penetrated in the utility [8]. On the other hand, the rising energy needs could not be fulfilled by the current weak grid supplied by the majority generation based on DG [7]. However, the intermittent characteristic of PV has major concern to be solved especially in isolated grid. Meanwhile, there are hundreds of units of DG that need either to get them hybridized with VRE or interconnect existing DG and VRE altogether in the system.

To prepare for the reliable system operation in the mentioned program above, pilot project in Sumba Island can be discussed since it has VRE penetration among most of the diesel generation. Thus, on grid PV system is currently being considered as an attractive, and clean option for isolated grid operations. This application is a means to reduce diesel fuel consumption. VRE that integrated with DG systems are currently being considered as an economic, and clean option for isolated microgrids, to offset oil fuel consumption by shifting generation from existing units; However, the security and stability of the system are a challenge from the increasing penetration VRE [6]. Meanwhile, Indonesia adopted European countries use electric frequency 50 Hz. If this electric frequency in the power network is not constant, the electrical equipment connected to it would have potential to get damage. So, the frequency needs to be maintained in a tolerable range. To keep this frequency stable and constant, it needs an automatic control scheme that can identify changes in frequency, then mitigate those changes by choosing which dispatch from DG has the lowest economical losses, which is known as the Automatic Dispatching System (ADS) [10].

The ADS was installed at the several DG units in eastern grid of Sumba Island since it is interconnected to the 1 MWp installed capacity of solar PV. Consequently, an isolated grid such as Sumba system has a potential risk of frequency instability. It is because the electricity sources are supplied by PV with intermittent characteristics. Under such conditions, implementing an efficient secondary controller through load frequency control (LFC) plays an important role in restoring power system stability [3].

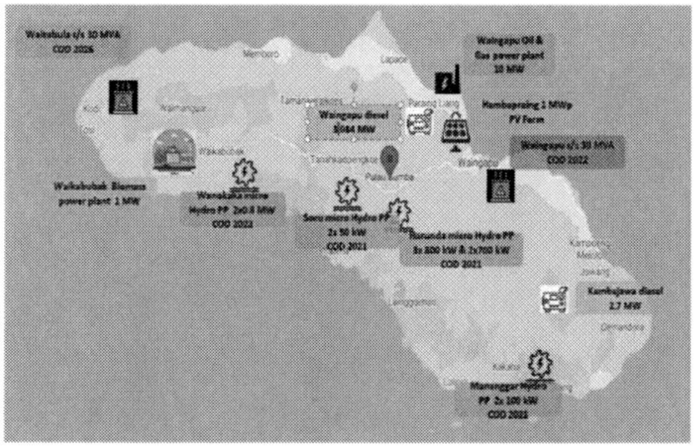

Figure 1. Sumba distribution grid and its planning expansion

Sumba island is in the southern part of Indonesia, which has population more than 250 thousand with the electricity demand less than 7 MW. Meanwhile, installed capacity of the total generation close to 16 MW, which majority 70 % supplied by diesel generation, and the others sourced from renewable energy, such as hydro, biomass and PV as indicated in Figure 1.

The load profile in Sumba has typical load profile characteristic which has low load in the daytime and increasing load demand in the evening. The Majority of typical load on the PLN East Sumba network is of a residential type with the characteristics of time in Figure 2. The peak-load 6.8 MW, and during, the daytime load, has parallel operation with the maximum PV generation week's cycle, which fluctuates between 3,7 MW – 4,8 MW.

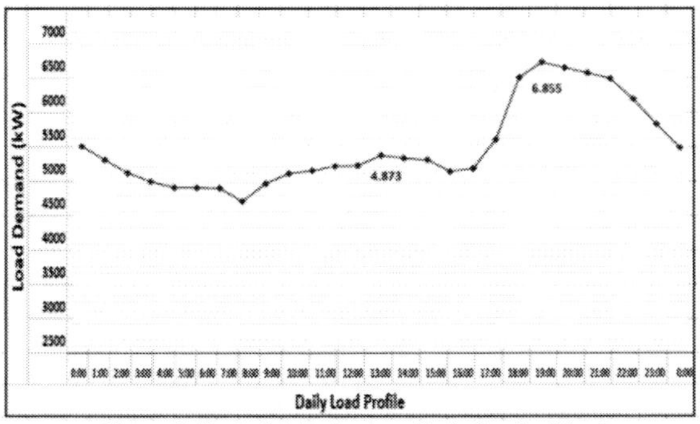

Figure 2. Typical daily load profile in Sumba Island

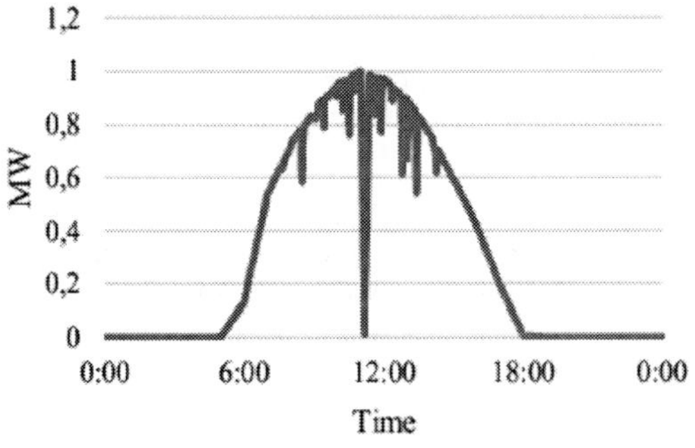

Figure 3. PV Output Power

Currently, Sumba Island has two main grids, eastern and western grid which the majority supply based on diesel fuel generation. East grid has PV generation with the installed capacity 1 MWp, which now represent 10 % of the capacity. Meanwhile, most of the diesel generations are operating in load following mode with manual dispatching. In addition to diesel, the Sumba electricity network is supplied with a Hambapraing PV with a capacity of 1 MWp which the output is varies with time movement as shown in Figure 3.

TABLE 1. POWER GENERATION COMPOSITION IN SUMBA

No	Owner	DG Unit Name	Installed Capacity (MW)
1	PLN	Cummins	1.2
2		Cummins QSK 1	0.7
3		Caterpillar	0.7
4		MTU	0.5
5		MAN 1	0.5
6		MAN 2	0.3
7		SWD DRO 1	0.3
8		SWD DRO 2	0.27
9		Yanmar	0.25
10		Deutz BA 1	0.25
11		Deutz BA 2	0.25
12		Deutz BA 3	0.25
	Total		5.47
13	Rent	Dossan 1	0.5
14		Dossan 2	0.5
15		Dossan 3	0.5
16		Dossan 4	0.5
17		Dossan 5	0.5
18		Dossan 6	0.5
19		MTU 10 V	0.4
20		MTU 12 V 1	0.5
21		MTU 12 V 2	0.5
22		MTU 12 V 3	0.5
23		MTU 12 V 4	0.5
24		MTU 12 V 5	0.5
	Total Rent		5.9
	Renewable Energy		
25	IPP	PV Farm	1
26	Local Gov	Biomass	1
27	IPP	Hydro	1
	Total Renewable		3

In TABLE 1 above, it can be seen that the PLN's DGs available on the isolated grid will be given the ability to respond to changes in load. In most cases, PLN's DG will be used to provide peak loads. PLN's generators can work with droop governor or with the isochronous governor. Thus, the isochronous governor regulates the valve/gate of the turbine to carry the frequency return to its nominal value [12]. So, the system with isochronous governor has the capability to restore the system frequency to 50 Hz after instability occurs. Then, the DG that works isochronous will be connected to ADS [3].

The objective of this paper is to evaluate the system performance of distribution grid using ADS and its impact in managing power quality and maintaining the penetration level of PV operation.

This paper is organized as follows, Section II presents ADS algorithm and control basic design of ADS which include simulation study of frequency path with combination of Droop and Isochronous governor. Section III describes ADS configuration that applied in the distribution grid. It also presents the development phase for smart grid and advantages using ADS in an isolated grid. Then, the output results of ADS installation are reported in Section IV, including the differences before and after ADS implementation based on the stability system performances.

II. METHOD

In this paper, ADS Algorithm is proposed as modeling to represent conceptual design of ADS used in the system. Hence, the model is implementing in the actual condition to justify whether the design is applicable and contribute benefit to the power quality in the system. In this study, ADS can be modeled by making an isochronous governor model that is connected to the PLN's generator in the system as showed in Figure 4. Then, the dynamic model of the governor is simulated using a simulation tool [3].

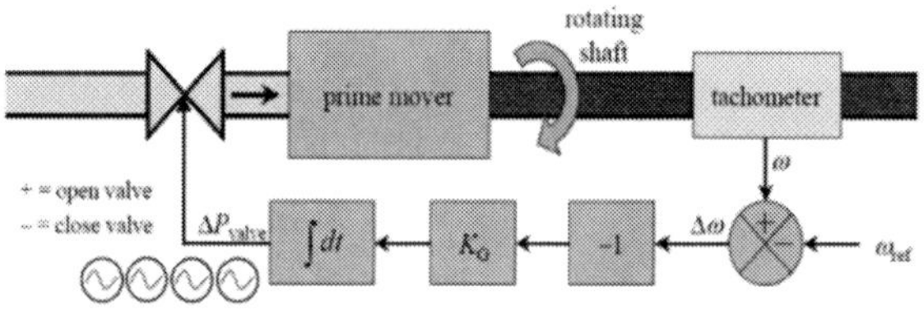

Figure 4. Control Basic Architecture Design of the ADS System

ADS is provided with an algorithm to turn the isochronous generator off and on when the load is either too low or too high. Hence, the generator needs to be switched off when the load is low because low loading will impact in inefficient energy conversion processes in its. So, the generator needs to be switched on when the load is getting high since the load will not get sufficient sources. This would lead to the collapse of the system frequency [3].

The proposed algorithm can be seen in the flow chart in Figure 5. Referring to table 1, the PLN has largest generator, DTG Cummins 1.2 MW, will be the generator with the most priority to continue running. Its priority to be switched on is ordered by the magnitude of its rating. Subsequently, the generator that has the lowest capacity, DTG Deutz BA 3 250 kW, will be the generator option with the most priority to be switched off. Hence, the lower the DG rating, the more frequently it will be either switched on or off. The DG with the largest installed capacity is prioritized to be switched off because if it is changed prioritized to turn it on and off, there would be a greater frequency swing than if the PLN DTG with the smallest rating is often turned on and off [3].

In general, ADS has automatically decided which generations have to be switched on, synchronized, loaded or stopped. Consequently, this solution can handle any type of power generation technology, as well as can be applied to any grid in any size. Meanwhile, solar PV and the diesel plant are developed at Kambajawa Distribution Station.

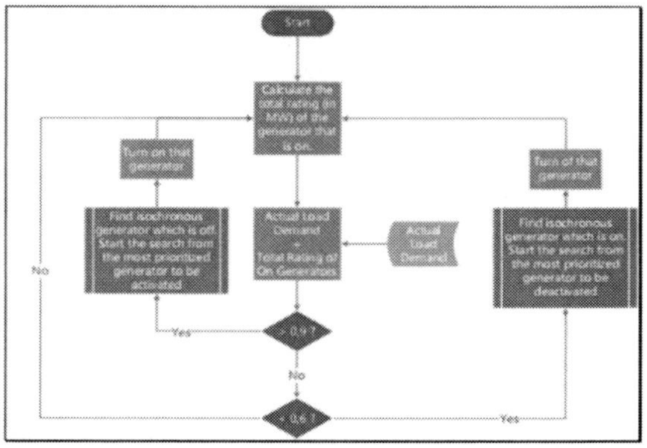

Figure 5. Algorithm of ADS

Current ADS reflects Multiple Generator status (on/off, synchronized or disconnect) and Multiple generator dispatch. Hence, ADS send power instruction, not just control signal. Thus, it reads the status of the grid and the generations in real-time instantaneously, and automatically instructs the diesel generations to dispatch certain actual power to stabilize the frequency, while at the same time, it maintain generations in operation within its optimal power settings.

Based on the simulation carried out by [3], the use case of 9 unit PLN's DG with primary controller, or Droop Governor, and 3 units with Isochronous governor is simulated to verify the output frequency within its in normal range (49,5 s.d 50,5 Hz). As shown in Figure 6 below, the simulation study indicated that frequency trajectory from the interconnection of Unit PLN's DG with Droop Governor and Unit PLN's DG with Isochronous Governor. The blue chart indicated Droop 5% with delay governor 20 s, while orange one showed Droop1,6% with delay 60 seconds.

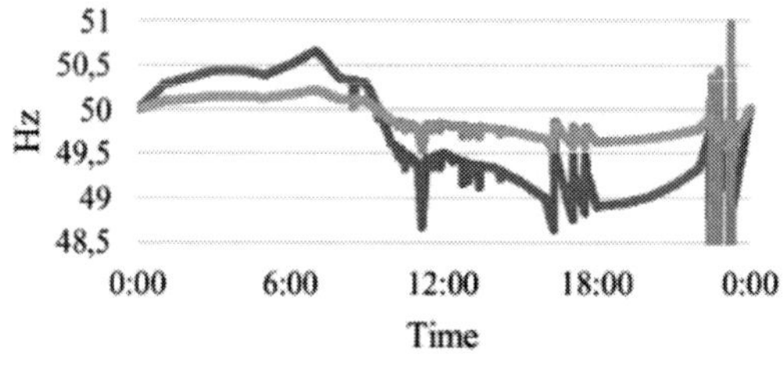

Figure 6. Frequency path with combination of Droop and Isochronous governor

Using ADS is expected that the isolated grid frequency can be adjusted in the tolerable range, without the need to implement an energy storage system [3].

III. ADS CONFIGURATION AND ITS PROPOSED DESIGN IN THE DISTRIBUTION GRID

Automated dispatching control system (ADS) is a control system tool which has function to monitor the system of the load flow in power network from various power generation types whether conventional and variable renewable energy, thus it capable to enter the electricity and energy market, as well as to accommodate the energy mixed. ADS is used to manage the various power supply characteristics capable of meeting the fluctuation of their load demands. On the other hand, ADS will be the flexible solution as the demand response in this distribution grid.

ADS have been widely implemented in any system voltage level system whether in medium voltages or high voltages. Meanwhile, the important role in the configuration of ADS, as the main direction of control system development, was the formation and implementation of SCADA, which collaborates information about the processes and centralize control. Thus, ADS is a type of automated process control which can control any type of power generation roles in the grid [2].

There are several methods that can be applied to improve Grid quality in Sumba system : implement ADS over East grid gensets, interconnect both grids (east and west grid) through high voltages lines, implement ADS over west grid power generation once upgraded, and develop the first operational smart grid power management [1].

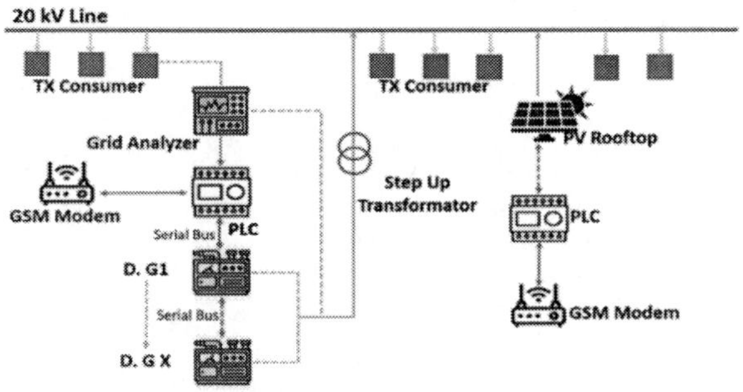

Figure 7. Conceptual Layout of Proposed System Components

Figure 7 describes the conceptual layout of the system. Depending on model, grid analyzer can read the grid from the LV bus (option 1) or from a consumer type step down transformer on the main HV outcome (option 2). Thus, the PLC can be replaced by a full size computer or a laptop computer, while GSM modems or more complex telecoms can be implemented for remote management and/or integrate with an active solar SCADA. Meanwhile, there are basic requirements to be implemented for preliminary phase in smart grid, which consists of [1] :

1. All flexible generators to be operated exclusively in full range load following mode, with minimum operation setting defined at 30% or lower and maximum operation setting defined at 80%
2. All flexible generators to have ECU's, electronic governors and Modbus and, optionally, TCP/IP protocols. Removal or redefinition of the now obsolete grid KPI of SFC; to be substituted by the actual international standard of liters per kWh delivered
3. Integrate them in the ADS/AGC system for automatic and sequential dispatching of the generators according to grid load behavior.

Implementation Phase for Diesel Reduction strategy and smart grid Introduction in Sumba Island as illustrated in Figure 8 was the upcoming flexibility solution for distribution grid improvement in Sumba Island since high grid losses and fuel consumption dominated the operational system, which disrupt the power quality and reliability of electricity sources to the customers [1].

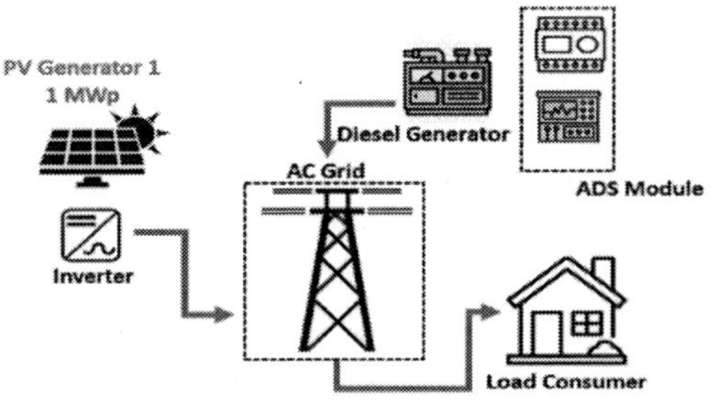

Figure 8. Pilot project outline for Diesel Reduction and PV integration

The development for smart grid can be executed by implementing the following systematic method as mentioned [1] :

1. Grid impact study, by doing the simulation study and modeling of the power system in existing Sumba grid that combined with the proposed generations, would be obtained the load flow analysis and the other electric output parameters.
2. Sizing and design of diesel buffering engine/engines, would be required to estimate the appropriate diesel capacity that are feasible to integrate with VRE. Hence, diesel sizing also compulsory to determine the ramping rate in the system
3. All diesel should be equipped with ECU/Modbus controls, and operated in stepped dispatching procedure
4. Sizing Solar PV up to 60% of diesel, one way to implement the high level penetration by integrating the PV share which is higher than diesel generation. This can be combined by applying ADS, as the hybrid controller unit that has function to control the flexible operation mode between PV and diesel. Subsequently, the ADS would match the synchronization between generation sources and the load demand.
5. Data acquisition units in all grid key nodes, data acquisition as the controlling and monitoring unit can be applied to assist the communication and remote

surveillance to ensure the smooth operation in the system.

Consequently, there are some advantages of using ADS in the isolated grids. During the test, the ADS has capable to control the grid frequency at stable rate by managing the operation of Kambajawa diesel power plant to meet the variability of power load/demand as well as the variability of 1 MWp solar PV power plant output. Thus, PLN will continue to observe the situation, especially during critical hours (peak hours or high ramping rate hours) to ensure whether the controllers and the generation power system balance power supply and power load if a nearby 1 MWp solar PV power plant supplies electricity intermittently due to weather condition [2].

In addition, ADS operation mode and its dispatch would not only guarantee the power quality and grid stability, but also it will support the induced load reduction from the Solar PV and other VRE that would be installed in the future. By allowing PLN to capitalize on the fuel usage savings and the cost spin-down effect of the incorporation of continuously lower cost VRE generation, which subsequently would lower PLN's operational cost as well as increase corporate profits.

IV. Output Results

The main issue in the distribution grid before ADS installation was the difficulty to maintain frequency stability in East Grid when the intermittency condition occurred as the result of fluctuation of output operation from the PV due to the PV soothing and solar irradiance changing.

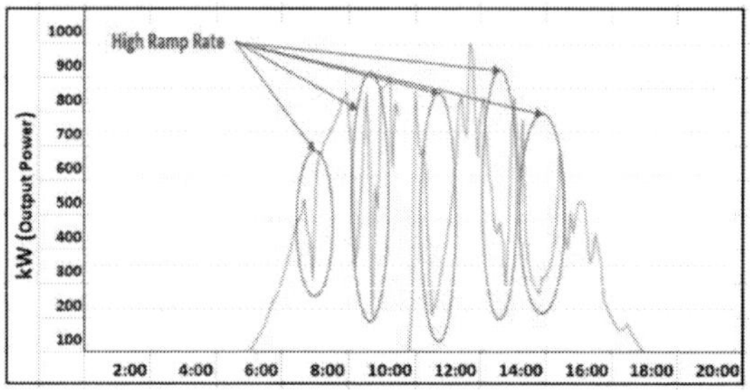

Figure 9. Typical electricity production of the Solar PV 1 MWp in East Sumba during cloud events.

The High Ramping Rate triggered the sudden changing of power frequency, which can impact the power quality and stability in distribution grids which is implied in Figure 9 where the PV output varied during cloud events.

There are preexisting condition before ADS integrated in the system, such as manual dispatching and wider range of grid frequency operation, constant power spillage by overpowering the grid, high thermal losses, and high-power leaking to neutral, high instability of the grid during operation hours of PV, and Black out/feeder disconnection due to under frequency.

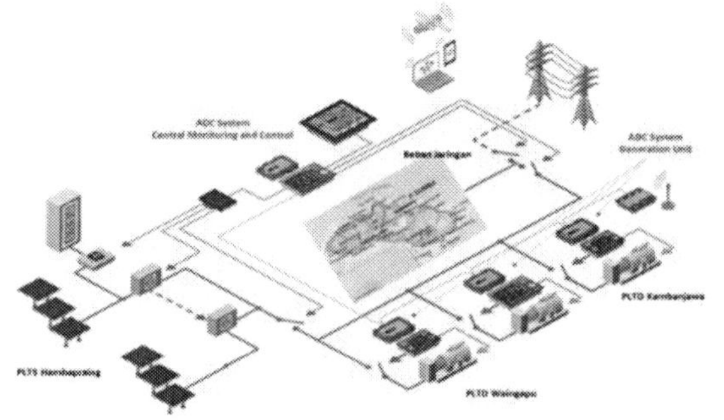

Figure 10 The proposed of ADS configuration in distribution grid

ADS system can control the complex mini grid operation that consist of PV and diesel generation as implied in Figure10, which is possible to integrate between many renewable energy sources with diesel generation by maintaining stable power quality with high efficiency and security level [1].

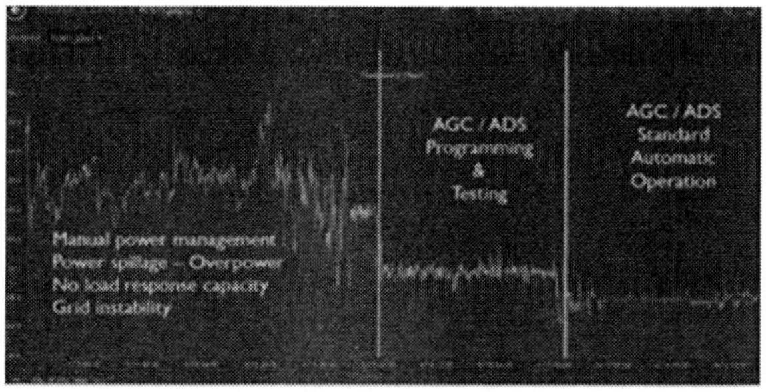

Figure 11. Typical electricity production as the impact of the Solar PV's operation

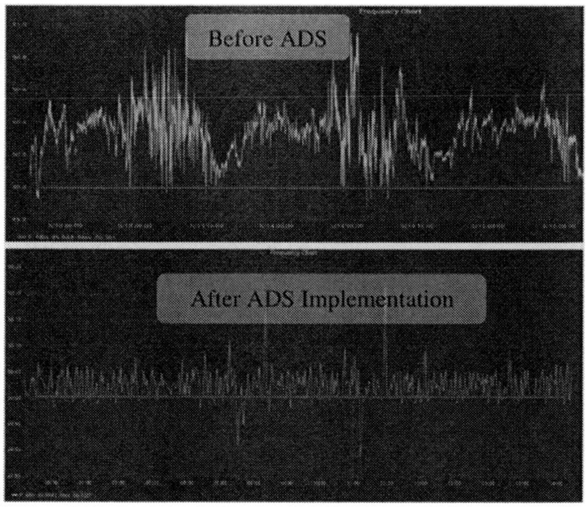

Figure 12. Typical electricity production in the system, before and after ADS installation

The graph displayed in figure 11 and 12 above described that ADS has primary roles to maintain the frequency stability in its nominal ranges by adjusting mechanical power in the power generations. The fig implied that before ADS installation, the frequency fluctuated above the tolerable limit. After ADS installation is applied, the frequency can be maintained within normal range (49,5<Hz<50,5). Yet, ADS utilization offers flexible operation since main devices can be modified as needed, as well as modifying its control algorithms in a good manner. Subsequently, the outcome emerges after ADS Installation which referring to the data diagram obtained that frequency system fluctuation is better adjusted within the nominal required range compared with the previous condition. In this use case, ADS can support distribution grid improvement by maintaining the stability of frequency in the system.

V. Conclusion and Future Works

1. The ADS implementation is possible to raise smart grid's operational performance, and provide reliability as well as quality power to their consumers, while minimizing the operation cost, which can be achieved by embracing the ADS installation as the first step to improve the power quality, thus minimizing losses in the distribution grid.
2. The development urgency of ADS in Sumba considers high generation cost, flexibility of system operation, and the potency of power factor correction. On other hand, ADS has proven capable of maintaining frequency stability in the system that contains intermittent power generation. Both the simulation, and actual measurement testing, it can maintain the stability frequency within normal range (49,5<Hz<50,5)
3. ADS implementation and grid monitoring system can be considered as the main part in integrating variable renewable energy into small power network, not only to compensate the fluctuations characteristics of VRE, but also to maintain grid stability and prevent the grid operations from abnormalities, such as over frequency, phase imbalance and frequency hunting.
4. The success of ADS installation, testing and operation boosted PLN's understanding and confidence to replicate the innovation to increase VRE integration in the upcoming de-dieselization program.

ACKNOWLEDGMENT

The author would like to express the gratitude to the joint working team between PLN and USAID ICED team who have great effort to develop and configure ADS implementation project in Sumba Island. Within the period 2018-2019. Subsequently, the ADS project obtained full support from the US Government. Thus, this smart grid project would be typical of pilot project in the future in deploying sustainability of renewable energy since this project can be potential solution to be expanded in de-dieselization program into more affordable electricity access as well as promoting clean energy environment.

References

[1] USAID - ICED II Team: 'The Summary of Grid Improvement in Kupang and Sumba ', May 2018, pp. 1–21

[2] USAID - ICED II Team: "Development of Automatic Dispatch Controller System and Grid Monitoring for East Sumba Grid ', Sharing Presentation, May 2019, pp. 1–46

[3] Bangun Primanta Holand, Haryanto Nanang "Design of Automatic Dispacth for Control System For Frequency Control of Isolated Electrical Grid" School of Electrical Engineering and Informatics, Institut Teknologi Bandung, Master Thesis, 2021

[4] T. Reber, K. Burman and B. Hirsch, "Sustainable Energy for Remote Indonesian Grids: Strategies to Accelerate Nationwide Deployment," Department of Energy by the National Renewable Energy Laboratory (NREL), Denver West Parkway, CO, 2016.

[5] H. Kanchev, F. Colas, V. Lazarov and B. Francois, "Emission Reduction and Economical Optimization of an Urban Microgrid Operation Including Dispatched PV-Based Active Generators," IEEE Transactions On Sustainable Energy, vol. 5, no.4, pp. 1397-1405, October 2014.

[6] W. Mendieta and C. A. Canizares, "Primary Frequency Control inIsolated Microgrids using Thermostatically Controllable Loads, "IEEE Transactions On Smart Grid, vol. 12, no. 1, pp. 1-13, 2020.

[7] A. Veldhuis and A. Reinders, "Potential and Cost-Effectiveness of OffGrid PV Systems in Indonesia," in 2014 IEEE 40th Photovoltaic Specialist Conference (PVSC), 2014

[8] M. Datta, T. Senjyu, A. Yona, T. Funabashi and C.-H. Kim, "A Coordinated Control Method for Leveling PV Output Power Fluctuations of PV–Diesel Hybrid Systems Connected to Isolated Power Utility," IEEE Transactions On Energy Conversion, vol. 21, no. 1, pp. 153-162, March 2009

[9] K. Zou, A. P. Agalgaonkar, K. M. Muttaqi and S. Perera, "An Analytical Approach for Reliability Evaluation," IEEE Transactions On Smart Grid, vol. V, no. 6, pp. 2657-2665, 2014.

[10] Johansson, Thomas & Phatwardhan, Anand and Nakicenovic, "Global Energy Assessment: Toward A Sustainable Future, Cambridge University Press, 2012

[11] Bangun Primanta Holand, "Automatic Dispacth for Control System Design" School of Electrical Engineering and Informatics, Institut Teknologi Bandung, Poster, 2020

[12] P. Kundur, Power System Stability and Control, McGraw-Hill, 1994

Implementation of a Fully-automated Optimized Fog-computing-based IoT-controlled PV Network

Adel M. AlWaqfi
Electr. Eng. Dept.
Al Hussein Technical University
Amman, Jordan
email: adelalwaqfy@gmail.com

Mohammad J. Abdel-Rahman
Data Sci. Dept., Princess Sumaya University for Technology
Electr. Eng. Dept., Al Hussein Technical University
Amman, Jordan
0000-0001-5788-6656
email: m.AbdelRahman@psut.edu.jo

Yunis A. Al-Qreenawi
Electr. Eng. Dept.
Al Hussein Technical University
Amman, Jordan
email: yalqreenawi@gmail.com

Zain AlHwaidy
Electr. Eng. Dept.
Al Hussein Technical University
Amman, Jordan
email: zainalhwaidy@gmail.com

Ahmad I. Abu-El-Haija
Electr. Eng. Dept.
Jordan University of Science and Technology
Ar Ramtha, Jordan
email: haija@just.edu.jo

***Abstract*—The Internet of Things (IoT) is revolutionizing almost every aspect of our lives. Smart grids are one example of this. Integrating IoT into various fields is a challenging task. In this paper, we present a detailed implementation of a fog-computing-based IoT system for monitoring and controlling a photovoltaic (PV) power network. As a case study, the implemented system is used to facilitate automatic energy routing within two PV systems feeding different loads. Our results demonstrate the ability of our IoT system to efficiently and automatically monitor and control a network of PV systems.**

Keywords- Photovoltaic (PV) systems; IoT; fog computing; automation.

I. INTRODUCTION

The Internet of Things (IoT) is a collection of technologies that grant us the ability of interfacing almost any physical object in the world, that is of interest in terms of its data and actions, to the digital world of computing, making it possible to monitor, analyze, and control mechanical and electrical systems everywhere, as well as driving insights about those connected things. The research and implementations in fields such as energy, healthcare, retail, transportation, and agriculture are continuously emerging and coming into reality in the form of smart gadgets, smart homes, smart factories, smart grids, self-driven cars, and smart cities [1][2].

In essence, the availability of real-time sensory data and sufficient computing resources constitutes the real power of IoT. However, the huge amount of data that will be generated in the process of digitizing the world requires deliberate design of the underlying infrastructures and technologies that are responsible for data delivery, processing, and storage. Here comes the concepts of cloud, fog, and edge computing into play [3]. Cloud computing can provide the necessary hardware and software for hosting, managing, and operating IoT solutions in the form of Infrastructure as a Service (IaaS), Platform as a Service (PaaS), and Software as a Service (SaaS). Nevertheless, edge and fog computing concepts are believed to be supplemental technologies that become necessary when the latency of delivering the solution is crucial or local (i.e., near the source of data) awareness of the solution is required.

Wireless Sensor Networks (WSNs) can be integrated into the Internet to move the currently connected networks from the local to the global domain [4][5], which essentially extends the landscape of IoT. Also, due to the fact that IoT applications are meant to be data-centric, there is a need for IoT and WSN communication protocols that are well-designed for this purpose. Existing application-layer IoT protocols in the literature are AMQP, CoAP, XMPP, MQTT, and HTTP. Other protocols that cover different layers of the OSI model are BLE, WiFi, Sigfox, NB-IoT, Zigbee, Zwave, and 6LoPWAN [6].

One of the most important applications of IoT is the Smart Grid (SG). SG is an electrical power network characterized by an infrastructure that makes it capable of sensing, communicating, monitoring, and controlling all parts of the network and addressing problems when occurring, unlike traditional networks [7]. In other words, a conventional grid can be transformed into a smart one with the help of IoT [8]. Renewable Energy (REs) sources such as wind and photovoltaic are considered the main and the most important generation sources relied upon in the smart microgrid. The stochasticity and unpredictability of the REs power generation make the smart grid nature potentially unstable. This brings the need for the IoT network to monitor the time-varying behavior of the generation and the load demand. All those challenges in the power system can be solved by integrating IoT. On top of that, comes the optimization layer that aims at minimizing the total cost of the microgrids by finding the optimal configuration between sources and loads based on real-time data provided by the smart sensors and reported by the IoT network.

Our Contributions—Our main contributions are:

- We present a detailed implementation of an IoT solution to monitor and control a PV system starting from the lowest level of physical components (i.e., the sensor node), going through the communication, all the way to the fog node setup.
- We evaluate the setup by running a simple optimization algorithm on top of the collected data.

Paper Organization—The rest of the paper is organized as follows. In Section II, we review the previous works in the literature on IoT implementations for smart grids. In Section III, we explain in detail the steps involved in implementing each component of our setup. Preliminary results are presented in Section IV. Finally, our conclusions are summarized in Section V.

II. LITERATURE REVIEW

The potential gains and challenges of leveraging technologies like AI, IoT, and 5G in smart grids were discussed in [9]. The potential improvements in the system's robustness, reliability, resiliency, and security are faced with challenges such as the unreliable wireless channels used in IoT. In this section, we highlight a few papers that considered implementing IoT networks for smart grids.

In [10], the authors focused on the WSN part of the IoT stack, designed and implemented a new WSN platform that consists of power-independent nodes. The nodes employ rechargeable batteries using embedded photovoltaic batteries harvesting the ambient light starting from 100 lux. The authors in [11] presented an implementation for monitoring regular domestic conditions, such as temperature and light intensity. They used a ZigBee-based WSN, where the gateway of the network is responsible for translating the ZigBee format of data into an Internet IPV6 format before reporting it to a MySQL database on a Windows-based server and displaying the results using PHP and JavaScript on a browser. A WiFi-based WSN is implemented in [12] for monitoring and controlling environmental, safety, and electrical parameters of a home. The authors supplemented their stack with an Android application instead of a web-based approach. An end-to-end, dynamic, and scalable solution was implemented in [13]. The authors in [13] assume that all IoT devices are WiFi-capable and can communicate using MQTT. They used an ESP8266 microcontroller and a cloud-hosted server with a MongoDB database to store the data.

Optimizing the energy production, consumption, and storage in smart grids over time helps save costs and reduce energy wastage. IoT systems play a key role in automating and interconnecting equipment. As discussed in [14], the authors used an energy management algorithm to match the required load with the generated energy from PV and wind energy systems.

To the best of our knowledge, this paper presents the first solution that (i) combines the locality and reliability of RF-based WSNs with the powerful edge of a gateway and MQTT communications and (ii) leverages the structured nature of SQL database to store non-sensory data along with the dynamic properties of a time-series database (that are suitable for IoT data) and harness this to create a base for developing smart grid solutions. Our paper provides an end-to-end solution that illustrates the whole cycle of IoT, from data generation to response and actuation, without missing the middle parts of provisioning WSN, channeling and storing data, and algorithm deployment.

III. IMPLEMENTATION SETUP

We consider a multi-microgrid system that is managed locally (adopting a fog computing paradigm) by an IoT-based control system. Our system consists of three main components: A power system, a control system, and a communication system.

The implemented setup is designed to represent the considered IoT-controlled smart grid on a small scale, to test the feasibility of applying control algorithms to the power system. A stand-alone multi-microgrid system is built, consisting of two PV arrays with their inverters and batteries. We integrated IoT capabilities into the system by using sensor nodes, an edge node, and a fog node. The complete system with its three main components, power, control, and communication, is illustrated in Figure 1. Our implementation aims to have the best assignment of connections between microgrids to achieve reliable and available systems with the minimum cost. The components of the setup are as follows:

A. Power System

Two stand-alone PV systems, each with an AC capacity of 5.5 kWp were constructed at Al Hussein Technical University (HTU), Amman, Jordan. Each system consists of:

- Three stings with two polycrystalline Jinko panels, with a rated power 315 W (model type JKM315P) connected in series, south oriented with a slope of 24 degrees.
- Four 12 V-DC NPP batteries connected in series.
- A 5.5 kWp rated power MUST stand-alone inverter with a charge controller (model type PV18-5548VHM).
- AC cables 2C $\times 2.5$ mm^2 (CU/PVC/PVC) connected from the inverters to the loads.
- As a control device, a solid-state relay (model SSR-40DA).

B. Sensor Nodes (IoT Devices)

A sensor node is a combination of a Microcontroller Unit (MCU), a group of sensors, and a means to communicate its data with other nodes in the network or with a gateway. It can be also called an IoT device or a thing in the scope of IoT. IoT devices are generally designed to be resource-constrained in terms of energy and computational power. Thus, the integration between the sensor nodes and the fog layer is preferred to be through an edge node, which is not resource-constrained, can perform complicated tasks, and is equipped with advanced interfaces. In this case, the complexity needed for the fog layer communication is moved from each sensor node to the edge node. In our setup, multiple Arduino UNO microcontrollers serve as the mind of the IoT devices used to interface various deployed sensors. In addition, the UNO is

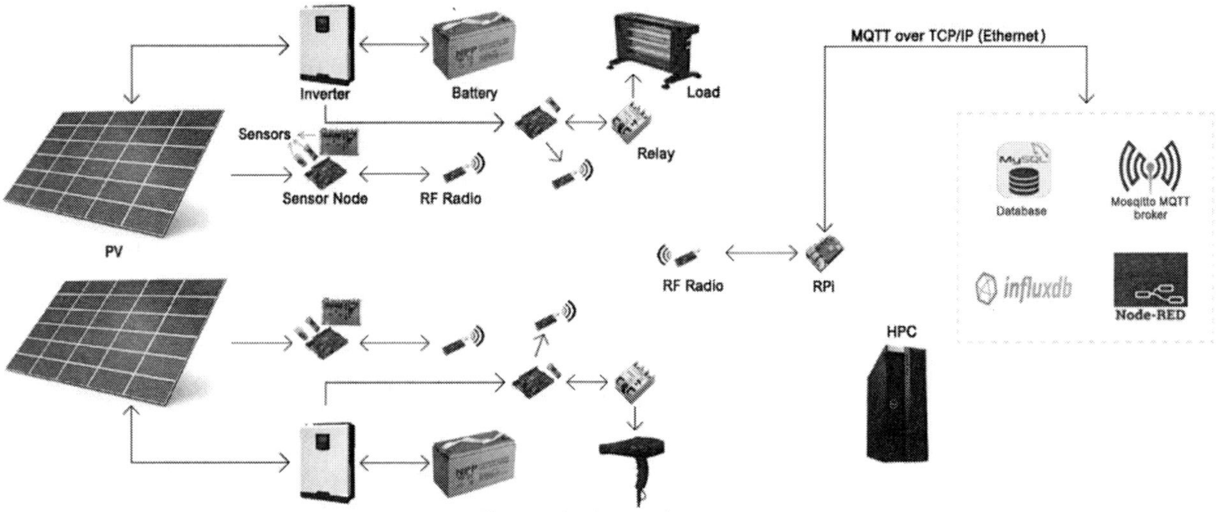

Figure 1. Implementation setup.

responsible for triggering the actuation process as a result of receiving commands in the IoT network. The microcontroller is based on an ATmega328 processor. It can support analog and digital sensors, as well as I2C and SPI sensor communication protocols. Usually, a node that performs only sensing tasks is called a sensor node or S node, whereas a node that performs both sensing and actuating tasks is called an SA node. We deploy a total of four nodes: three S nodes and one SA node. Two of the S nodes are used to read data from each inverter and one is connected to the load lines to measure their drawn currents. The SA node is used to control the assignment of sources to loads when needed.

C. Sensors and Actuators

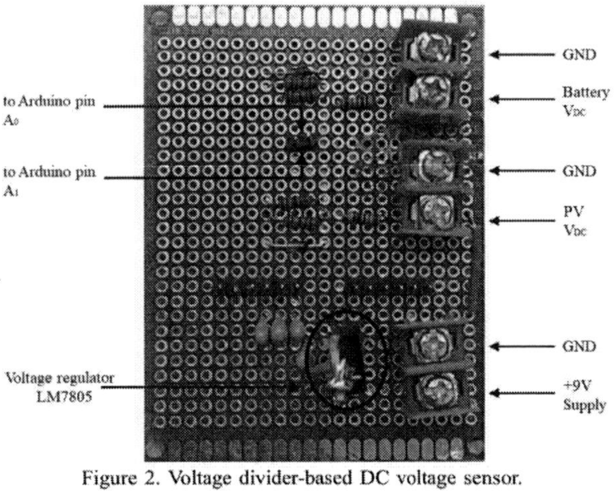

Figure 2. Voltage divider-based DC voltage sensor.

Multiple off-the-shelf sensors are used to comprise different sensor nodes in our system. They are used to measure quantities of interest for our use case. Those sensors are for measuring the electrical current and voltage of the DC output of the PV cells, the DC voltage of the batteries, and the AC output of the inverters.

For the current, we used ACS712 to measure the DC current output of the PV cells, the DC charging and discharging currents of the batteries, and the AC current output of the inverters. To measure the DC voltages of the PV output and the batteries, we designed a simple voltage divider-based sensor, shown in Figure 2. The sensor takes a DC input up to 50 V, converts it linearly into the 5 V range that the microcontroller can understand, and computes the voltage back. It uses the LM7805 linear voltage regulator to generate the needed reference voltage. For measuring the AC voltage, we used a transformer-based sensor, ZMPT101B. All the sensors have an analog output with a certain linear transfer function for a given range, which should be matched with the expected values to be measured. For actuating, we used solid-state relays (SSR-40DA) to control the contribution of each source (inverter) to each load.

D. Wireless Sensor Network

The communication inside the WSN is established using a wireless communication transceiver module nRF24L01+. The transceiver module is designed to operate in the worldwide ISM frequency band, and it uses GFSK modulation for data transmission. The center operating frequency is 2.4 GHz with 1 MHz channel bandwidth. With its reconfigurable transfer data rates (i.e., 250 Kbps, 1 Mbps, and 2 Mbps), it can serve the purpose of local wireless communication in the WSN. For an IoT application, 250 Kbps is sufficient to achieve the required performance. Moreover, the module comes with an implemented tree-topology network, in which using a single frequency channel out of the 125 available, can theoretically produce a network of up to 3,125 nodes. In Figure 3, we show the organization of the sensor nodes utilizing the tree topology, where each node is associated with a hexadecimal address to define its existence in the WSN. At the top of the network resides the gateway with an address of 00. Other sensor nodes are then assigned different addresses according to the tree topology. There are five depth levels, each can support up to five nodes. For example, the first level contains the nodes $01, 02, 03, 04$, and 05. Under node 01, other five nodes can exist with the addresses $011, 021, 031, 041$, and 051. Message Queuing Telemetry Transfer (MQTT) is an IoT protocol for transferring data in the application layer and was originally developed by IBM. MQTT is a lightweight and flexible asynchronous protocol, which means that it can be implemented on heavily-constrained hardware and limited-bandwidth networks. In MQTT, a subscriber is a client that needs to communicate with other clients. The broker in the protocol serves as the central entity that manages all connections and pipelines between clients based on the publish/subscribe model. Compared to other IoT protocols, such as CoAP, XMPP, and AMQP, MQTT is more suitable for resource-constrained environments. Also, in MQTT, the broker hides the complexity of communication rather than being on the side of the clients [15][16]. We use MQTT for communication between the edge node and the fog node.

The deployed nodes, shown in Figure 3, can be explained in details as follows:

- **Node 01 (PV1) and Node 02 (PV2)**: Each contains a UNO MCU, an LM7805, three ACS712, a DC voltage-divider sensor, a ZMPT101, and an nRF24L01+.
- **Node 03 (Loads)**: Contains a UNO MCU, four ACS712, and an nRF24L01+.
- **Node 04 (Actuator)**: Contains a UNO MCU, eight SSR-40DA, and an nRF24L01+.

Note that we have four electrical paths, one from each source to each load. Two relays need to be deployed for each path, one for the line and the other for the neutral. Hence, we have a total of $4 \times 2 = 8$ relays in Node 04. When a path is opened or closed, the two relays should be actuated at the same time to ensure proper flow of power and avoid problems.

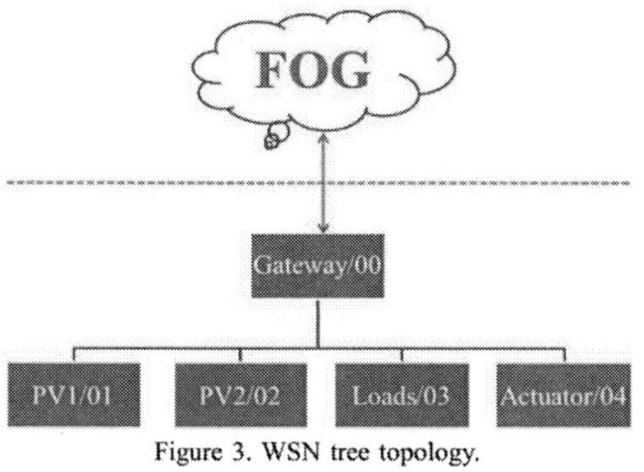

Figure 3. WSN tree topology.

E. Edge Node (Gateway)

Raspberry Pi (RPi) 4 is used to implement an MQTT client that bridges the local network of IoT devices with the MQTT broker running on the fog node. RPi 4 comes with a Quad-core Cortex-A72 (ARM v8) 64-bit CPU, 2GB of SDRAM, support of IEEE 802.11ac WiFi, Bluetooth 5.0 (BLE), and a Gigabit Ethernet interface. With these specifications, the RPi can easily function as a gateway in the IoT network, where multiple IoT devices constituting the WSN are able to deliver their data to the fog layer of the network using the aforementioned wireless transceivers. In addition, the RPi can be used, if needed, to process the data before delivering it to the fog layer, utilizing the edge computing concept.

F. Fog Node

The setup of this node can be divided into two parts: hardware and software.

1) Hardware

A High-Performance Computer (HPC) in our lab at HTU is used as the IoT fog node with an Intel(R) Xeon(R) Gold 6130 CPU running at 2.1 GHz and 64 GB of RAM. It is the infrastructure over which the application side of the IoT solution resides.

2) Software

The software setup depends on the desired functionality of the IoT solution. Basically, on top of an operating system, multiple software components are used to create an IoT platform. The heart of every IoT platform is the communication channel or broker and the database. Eclipse Mosquitto MQTT Broker is an open-source implementation broker for the MQTT protocol versions 5.0, 3.1.1, and 3.1. It is the broker we implement on our fog node for the purpose of application-layer communication. Also, we use Node-RED as the middleware that facilitates the broker-backend and broker-frontend connections to other parts of the solution. It is an event-driven programming tool for creating backend applications built on top of the Node.js framework. It is a very powerful tool for the development of IoT solutions as it needs an extremely small amount of code writing to achieve a lot of IoT-related tasks in the software stack with its user-friendly interface. An example depicting this simplicity is shown in Figure 4, where we set up the necessary nodes for debugging incoming and outgoing MQTT payloads as well as inserting the results into InfluxDB, which is the time-series database we deployed for storing the real-time data. Compared to conventional structured databases such as SQL, InfluxDB is schemaless and more suitable for real-time query to achieve data visualization or additional processing, data analytics, and machine learning tasks. A powerful aspect of using influxDB is the lightweight scripting language, which is Flux, that is used not only to query, but also aggregate, and manipulate the data with a very rich set of mathematical tools available with it. Moreover, Flux can be used to do this in real-time by utilizing the Tasks feature that can be deployed to schedule repeated and necessary patterns of computation on raw data to enrich it or prepare it for another phase of computing. We also set up a MySQL database, but it is used only to store information about the application in order to build the user interface (i.e., device ID, device name, etc.). InfluxDB comes with an interface that allows the user to define and configure data sources, explore the content of the buckets (i.e., databases), build dashboards with rich types of widgets, create scheduled tasks to be run on the streams of data, and set

threshold alert endpoints to monitor the connected IoT data. Figure 5 shows a dashboard we built to display various sensed parameters of one of our connected inverters. In the dashboard, we include two widgets per displayed value, one for graphing it and the other is a single-status widget that displays its latest state. Starting at the top from left to right are the RMS value of the AC output voltage, the DC voltage of the battery, and the DC voltage of the PV panel. Similarly, at the bottom are the output current, battery current, and the PV current. The previously mentioned components are all installed and configured on the HPC to build the IoT platform. Using the MQTT protocol, clients can either publish messages into a topic name or subscribe to a topic name to receive messages published by other clients or do both. To implement that concept, we used multiple topic names in the hierarchy form of $typeofData/Location/Gateway/Sensor. The topic names are:
$state/HTU/WiNS/Sensor-Node-Name
$command/HTU/WiNS/Sensor-Node-Name
$connected/HTU/WiNS

Topic names can be anything and can be organized into multiple levels by using the forward slash separator. The first topic is used to publish sensory data collected from each sensor node to the fog node. The second one is used to publish commands from the fog node to control the sensor nodes in the control system part of the setup. The third command is used to publish data about the connection and disconnection of the edge node to be able to track the connectivity state of the edge node.

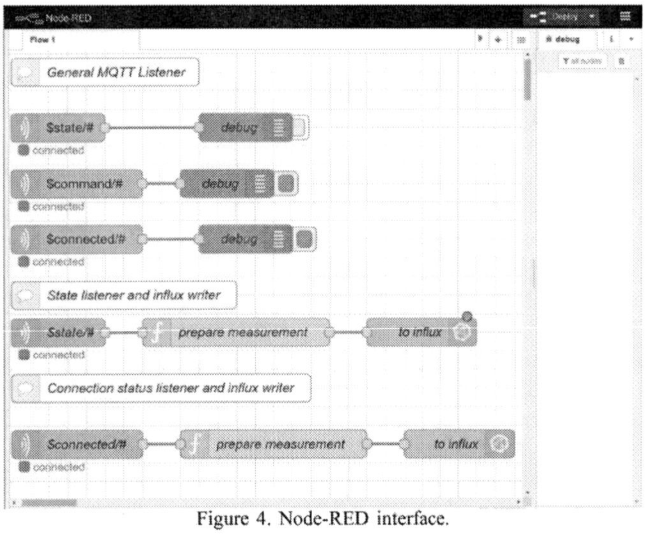

Figure 4. Node-RED interface.

IV. PERFORMANCE EVALUATION

The experimental setup, load swapping algorithm, and experimental results are presented in the following subsections. The codes used to program the WSN and implement the control algorithm in this section can be found at [17].

A. Experimental Setup

To validate the IoT system that we implemented, we used a simple algorithm for load swapping. A single source is expected to cover two different adjustable loads. The algorithm takes the sensed battery current and load current as inputs and gives the states of two switches controlling the state of each load as output. It compares the input values to predefined thresholds to determine if the connected load can be served during the targeted period. If not, it sends a command to actuate load swapping. In other words, the first load is turned off and the second is turned on. The process repeats every T seconds. We emulate the change of a load by adjusting the value of each load manually and waiting for the system to respond to the changes.

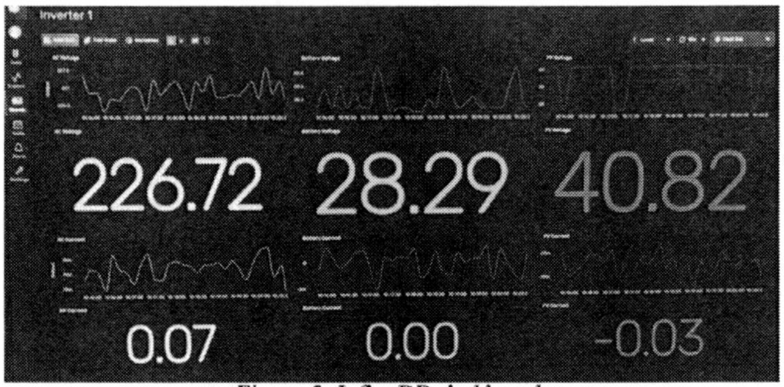

Figure 5. InfluxDB dashboard.

Input: I_b, I_o, $I_{\text{th},b}$, $I_{\text{th},o}$, τ_{delay}, and N.
Output: x_l, $\forall l \in \mathcal{L}$.
for $i = 1 : N$ **do**
  Set $x_0 = 1$ and unset $x_1 = 0$.
  Wait for τ_{delay}.
  Query latest I_b and I_o from database.
  while $I_b \leq I_{\text{th},b}$ **or** $I_o \leq I_{\text{th},o}$ **do**
    Wait for τ_{delay}.
    Query latest I_b and I_o from database.
  end while
  Unset $x_0 = 0$ and set $x_1 = 1$.
end for

Figure 6. Load swapping algorithm.

B. Load Swapping Algorithm

Let I_b and I_o be the battery current and the output current of the system, respectively. We define $I_{\text{th},b}$ and $I_{\text{th},o}$ as the battery current and the output current thresholds, respectively. Let $\mathcal{L} = \{1, 2, \ldots, L\}$ be the set of considered loads and let x_l be a binary decision variable. x_l equals one if load l is driven by the inverter, and it equals zero otherwise. In our case, $L = 2$ (i.e., we have two loads). Let τ_{delay} be the time difference between subsequent checks of conditions. Using the introduced notation, our load swapping algorithm for N time slots can be summarized in Algorithm 1 in Figure 6.

C. Experimental Results

We prepared our loads and ran the Python-based code that implement the algorithm and collected the following data in Figure 7 and 8. Figure 7 shows the changes in the states of load one and load two switches actuated from the fog node whereas Figure 8 captures the corresponding variations to the

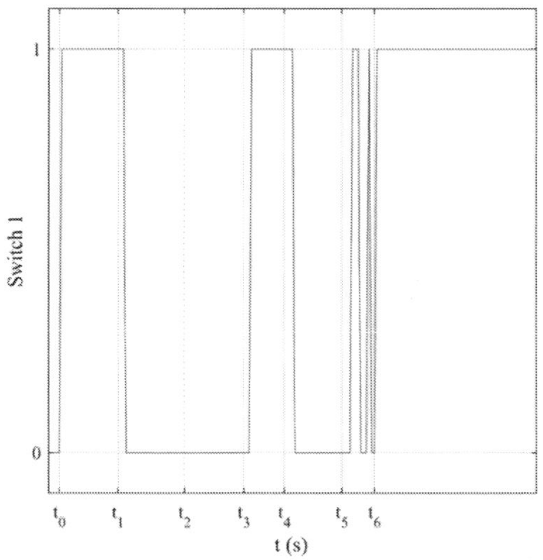

(a) Switch one.

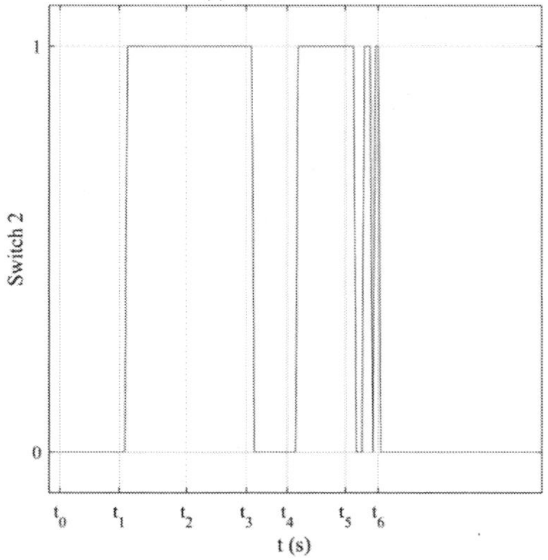

(b) Switch two.

Figure 7. The state of the switches during the testing period.

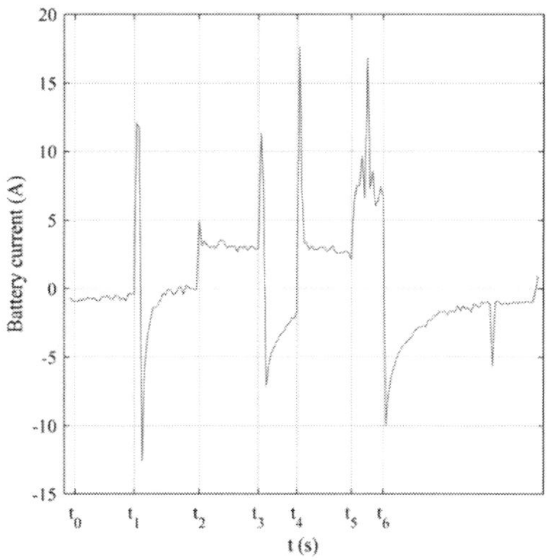

(a) Battery current (Ampere DC).

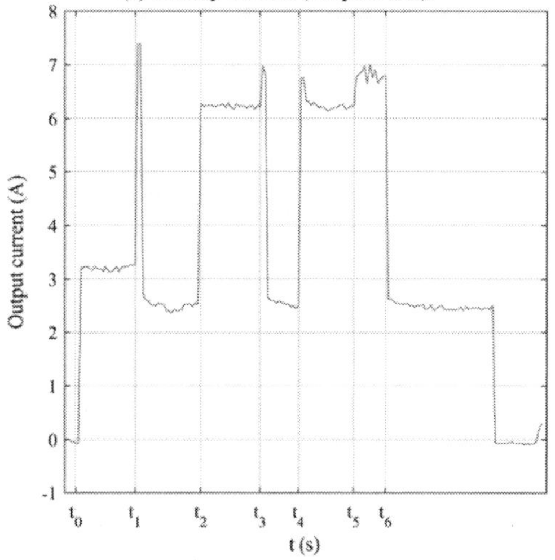

(b) Load current (Ampere RMS).

Figure 8. Current values during the testing period.

sensed battery and load currents. The scenarios start where both switch 1 and switch 2 are zero. Then, the fog node sends a $S_2S_1 = 01$ command to drive load 1 on. This action occurred at time t_0, as shown in the figures. After t_0, an increase in the load current is captured by the sensor. At the same moment, the battery current is negative which indicates that the battery is in the charging state. To push the system into using the battery we adjusted load 1 at t_1 where spikes in both the load current and the battery current are captured. This action pushed the currents beyond the threshold that we configured (5 A for each) which accordingly triggered the command that swapped the loads (i.e., $S_2S_1 = 10$). The figures confirm this and show that following this moment the switches flipped and the currents went back below the threshold. At t_2, we adjusted load 2 to be above the threshold. However, this time the battery contribution did not exceed the threshold which in turn kept everything as is in terms of the switches' states, as expected from the algorithm. Another increase of load 2 was triggered at t_3, which this time made the condition to be true again and the algorithm responded by swapping the loads as illustrated in Figure 7. t_4 shows another swap triggered by the same event. After this moment, specifically between t_5 and t_6, we fixed both loads at a heavy consumption state (the one that exceeds the thresholds). This action can be confirmed by the readings of currents during this period. The algorithm responded as expected by sending swapping commands repeatedly until the end of this period (at t_6) when we lowered load 1 to stabilize the switches eventually at $S_2S_1 = 01$.

V. CONCLUSIONS AND FUTURE WORK

We showed how each component of the IoT stack is implemented and how they are integrated with each other to collectively serve the expected purpose of the IoT, that is data collection, drawing insights, and operational improvements into the connected systems. Solving the challenge of integrating IoT into current systems will open the door for exploiting the collected data to create various algorithms and optimizations in IoT-based systems. We demonstrated that by implementing a load swapping algorithm on a fog-based IoT system to control PV systems. Despite being a simple threshold-based algorithm, it served the purpose of validating the challenging IoT setup that is composed of a multitude of parts and technologies.

Other major and crucial problems in similar systems can be solved using the same setup we implemented. Examples of such problems include resource allocation, load scheduling, energy cost optimization, and energy routing. Our future work will utilize the system built during this work to perform more complicated tasks, such as energy routing based on different objectives and constraints, utilizing the true power of IoT, fog computing, and time-series databases.

ACKNOWLEDGMENT

This work was supported by the Jordan Scientific Research and Innovation Support Fund (SRISF) (grant # ICT/1/9/2018). Any opinions, findings, conclusions, or recommendations expressed in this paper are those of the author(s) and do not necessarily reflect the views of Jordan SRISF.

REFERENCES

[1] A. Dawood, "Internet of things (IoT) and its applications: A survey," *Int. J. Comput. Appl.*, vol. 175, pp. 975–8887, Sep. 2020.

[2] S. H. Shah and I. Yaqoob, "A survey: Internet of things (IoT) technologies, applications and challenges," in *Proc. IEEE Smart Energy Grid Eng. (SEGE)*, 2016, pp. 381–385.

[3] B. Omoniwa, R. Hussain, M. A. Javed, S. H. Bouk, and S. A. Malik, "Fog/edge computing-based IoT (FECIoT): Architecture, applications, and research issues," *IEEE Internet Things J.*, vol. 6, no. 3, pp. 4118–4149, Jun. 2019.

[4] M. Ruiz, E. Álvarez, A. Serrano, and E. Garcia, "The convergence between wireless sensor networks and the Internet of things; challenges and perspectives: A survey," *IEEE Latin Amer. Trans.*, vol. 14, no. 10, pp. 4249–4254, Oct. 2016.

[5] N. Khalil, M. R. Abid, D. Benhaddou, and M. Gerndt, "Wireless sensors networks for Internet of things," in *Proc. IEEE Int. Conf. Intell. Sensors, Sensor Netw. and Inf. Proc. (ISSNIP)*, Apr. 2014, pp. 1–6.

[6] A. Amjad, F. Azam, M. W. Anwar, and W. H. Butt, "A systematic review on the data interoperability of application layer protocols in industrial IoT," *IEEE Access*, vol. 9, pp. 96 528–96 545, July 2021.

[7] B. M. Zakaria, A. S. Tag ElDien, and M. S. Elkholy, "Development of smart grid system," in *Proc. Int. Conf. Comput. Eng. and Syst. (ICCES)*, Dec. 2020, pp. 1–5.

[8] A. N. Pramudhita, R. A. Asmara, I. Siradjuddin, and E. Rohadi, "Internet of things integration in smart grid," in *Proc. Int. Conf. Appl. Sci. Technol. (iCAST)*, 2018, pp. 718–722.

[9] E. Esenogho, K. Djouani, and A. M. Kurien, "Integrating artificial intelligence Internet of things and 5G for next-generation smartgrid: A survey of trends challenges and prospect," *IEEE Access*, vol. 10, pp. 4794–4831, Jan, 2022.

[10] C. S. Abella *et al.*, "Autonomous energy-efficient wireless sensor network platform for home/office automation," *IEEE Sensors J.*, vol. 19, no. 9, pp. 3501–3512, May 2019.

[11] S. D. T. Kelly, N. K. Suryadevara, and S. C. Mukhopadhyay, "Towards the implementation of IoT for environmental condition monitoring in homes," *IEEE Sensors J.*, vol. 13, no. 10, pp. 3846–3853, Oct. 2013.

[12] N. Vikram, K. Harish, M. Nihaal, R. Umesh, and S. A. A. Kumar, "A low cost home automation system using Wi-Fi based wireless sensor network incorporating Internet of things (IoT)," in *Proc. IEEE Int. Adv. Comput. Conf. (IACC)*, Jan. 2017, pp. 174–178.

[13] J. Lohokare, R. Dani, A. Rajurkar, and A. Apte, "An IoT ecosystem for the implementation of scalable wireless home automation systems at smart city level," in *Proc. IEEE TENCON Conf.*, Nov. 2017, pp. 1503–1508.

[14] M. Kumar, A. F. Minai, A. A. Khan, and S. Kumar, "IoT based energy management system for smart grid," in *Int. Conf. Adv. Comput., Commun. Mater. (ICACCM)*, Aug. 2020, pp. 121–125.

[15] I. Heđi, I. Špeh, and A. Šarabok, "IoT network protocols comparison for the purpose of IoT constrained networks," in *Proc. IEEE Int. Conv. Inf. and Commun. Tech., Electron. and Microelectronics (MIPRO)*, May 2017, pp. 501–505.

[16] N. Naik, "Choice of effective messaging protocols for IoT systems: MQTT, CoAP, AMQP and HTTP," in *Proc. IEEE Int. Syst. Eng. Symp. (ISSE)*, Oct. 2017, pp. 1–7.

[17] A. M. AlWaqfi, *Implementation Codes*, https://github.com/Adel-Alwaqfi/Implementation-of-a-Fully-automated-Optimized-Fog-computing-based-IoT-controlled-PV-Network.

Spatial Analysis to Identify the Need for Additional EV Charging

Vivian Sultan, Benjamin Pezzillo, Roxana Cuevas, Solange Ruiz, Jinhua Tang, Amanda Acosta
College of Business and Economics
California State University
Los Angeles, USA
email: vsultan3@calstatela.edu

Abstract—**Los Angeles County's electric-vehicle (EV) charging-station infrastructure is growing rapidly due to increased environmental awareness. Implementing city-wide infrastructure is costly, and funding is limited. Spatial analysis can offer useful insights by visualizing existing EV charging-station infrastructure along with several other factors to determine where and how many EV charging stations are needed. Uneven EV charging-station availability complicates consumers' decisions to switch to driving EV. This analysis examines the uneven distribution of EV charging stations, the demand for more charging stations, and the EV-to-charger ratio to fill in the spatial discrepancies with charging stations in Los Angeles County.**

Keywords—Electric vehicles; charging stations; Spatial analysis; environmental awareness.

I. Introduction

For most EV drivers, switching to driving EVs has not been simple. However, the difficulties drivers face are not uniform, primarily due to affluence, population, and unequally distributed charging stations throughout Los Angeles County. To help spread EV driving, our analysis report suggests distributing more charging stations, preferably quick and vehicle-brand-neutral charging stations, in areas with spatial disparities, particularly in lower-income communities and areas where EV ownership is high in comparison to EV-charging-station availability. This would help smooth the transition for new EV drivers and reduce range anxiety. The same goal should also be applied to other communities. This is an important study because Los Angeles is among the top cities in California making the switch to EVs.

California has rapidly increased the number of zero-emission vehicles in the state, 1,300% in six years and from 25,000 in 2012 to more than 350,000 today. California continues to lead this trend as it has the highest EV penetration in the country, 40% of all EVs in the United States. To promote EV use and combat the lack of EV charger availability, California enacted Executive Order B-48-18. The order calls for "all State entities [to] work with the private sector and all appropriate levels of government to put at least 5 million zero-emission vehicles on California roads by 2030" and "all appropriate levels of government to spur the construction and installation of 200 hydrogen fueling stations and 250,000 zero-emission vehicle chargers, including 10,000 direct current fast chargers, by 2025" [1].

The Biden-Harris Electric Vehicle Charging Action Plan was created to push American leadership on using such clean cars as EVs. President Biden set an EV market share goal of 50% in the United States by 2030 and further explained the issue with the current EV charging network of over 100,000 public chargers as they operate "with different plug types, payment options, data availability, and hardware hookups." The Bipartisan Infrastructure Law "includes $5 billion in formula funding for states with a goal to build a national charging network" and "provides $2.5 billion for communities and corridors through a competitive grant program that will support innovative approaches and ensure that charger deployment meets Administration priorities such as supporting rural charging, improving local air quality and increasing EV charging access in disadvantaged communities." Moving forward, using the EV Charging Action Plan will establish a more uniform approach to EV charger accessibility while providing greater convenience for customers and offering increased confidence for the industry, which should promote the use of EVs in return [2].

In Los Angeles, the EV-to-charger ratio is currently 5.6. By 2030, this ratio is expected to grow to 37.4 as the EV industry is projected to grow immensely. Under the Biden administrations Plan for EV Infrastructure, the ratio goal is 18.7. Our research provides a means to reach this goal.

The paper is structured as follows. Section II reviews the literature on this topic. Section III presents the problem. Section IV presents the data. Section V presents the system used and the methodology. Section VI discusses the results and Section VII discusses the project's limitations. Section VIII concludes.

II. Literature Review

One factor we analyzed is charger locations in lower income communities. Identifying and prioritizing EV planning and extending the economic and environmental benefits to those who need them most is vital, with the focus on places with lower incomes and high environmental pollutants. Bui et al. [3] presents a figure showing the west side of Los Angeles with more cumulative EV registrations than the northeast and south. After adding our EV-ownership data, the gaps clearly represented EV ownership disparities. Peer-to-Peer Car Charging (P2C2) offers a scalable method for charging EVs that reduces the requirement for complex charging infrastructure. The idea is to work with a cloud-based control system to coordinate EVs so that they can share chargers while enroute. This would reduce the barriers and limitations consumers face in transitioning to EVs without having more local EV charging stations [4].

In contrast, our proposal is to carefully place more fast-charging stations. We start by outlining the three tiers of equipment used to recharge battery EVs. To help in the deployment of the variety of vehicles being offered, all three levels of charging infrastructure, from basic, low-cost Level 1 to DC fast chargers, should be built. "DC fast chargers will be important to drivers who need to quickly recharge their depleted batteries. Using fast chargers, most vehicles will recharge up to 80 percent of capacity in a ½ hour or less" [5]. This will help reduce range anxiety. A study in Ireland conducted a sensitivity analysis and discovered that the overall EV cost is highly sensitive to the daily availability of EV charging and the number of charging stations [6]. As an alternative to using big data, [7] used a forecasting model to calculate the demand for charging electric vehicles. The report used a decision tree for classification, a relational analysis to find influential elements, and cluster analysis to categorize traffic patterns [7]. This proposal would enable system engineers to foresee the need for EV charging based on historical traffic and weather data. Abo-Khalil et al. [8] evaluated the impact of random factors connected to vehicles' prior driving distances and the time at which they connected to the electrical grid to adjust the capacity of charging stations. Another issue is that neighborhoods in different counties of California with a large Black or Hispanic populations have less access to public EV chargers. The results from [8] show that the distance off the highway or freeway is negatively correlated with charging-station density. Los Angeles County was not included in Chih-Wei and Fingerman's analysis [9]. In Orange County, [10] created a machine-learning framework to investigate spatial discrepancies in EV-charging-station deployments using a predictive methodology. "The first was to pinpoint the crucial socioeconomic variables, and the second was to use these elements plus ground truth information from current charging station placements to estimate future EV charging station density using machine learning techniques at various spatial scales and compare their predictive performance to determine the best spatial resolution." Optimally sited charging stations are required for long-distance demand in heavily traveled areas like Los Angeles and New York. He et al. [11] analyzed long-distance travel data using spatial analysis and an algorithm demonstrating that a 100-mile range is required to prevent problems. Our study will determine the quantity and placement of fast charging stations for various scenarios, better planning, and more environmentally friendly transportation.

In the past, analyses have assessed the economic benefits of newly built public charging stations. These analyses have considered charging revenues and costs, land-rental costs, and investment costs. Charging revenues and costs are calculated based on electricity price, the cost of the selected electric vehicle supply equipment, and charging demand covered by the public charging station [12]. A common question in these studies is how much charging infrastructure is needed in each area and where charging infrastructure should be built. These question is crucial because these decisions are associated with large investments and have a long-term influence on adapting electric mobility [13]. We set out to answer them in our research. Careful studies have also been undertaken to identify the communities that EV charging station implementation tends to overlook. However, the measurement of "underserved" communities shifts according to the type of research and, therefore, could have many definitions. Zhou et al. [14], for example, looked at communities that experience high transportation energy costs, and high exposure to pollutants, which lead to public health issues and limited access to clean and reliable transportation. These factors were used to categorize underserved community needs. One point is certain: Lower-income EV drivers are more likely to rely on public charging [15], in large part because private EV charging stations are limited to homeowners. However, it is critical to implement a steady plan for EV infrastructure that not only supports current EV owners' needs, but is able to accommodate future EV owners. EV ownership requires access to both public- and home-charging infrastructure so that they can feel confident in transitioning to EV ownership without fear that their driving behavior will be curtailed due to refueling limitations [16].

As society moves towards using EVs, suppliers must work to meet consumer needs and demands. To sustain EV commuting, a well-running and coordinated electric-charger infrastructure must meet the charging demand. Aside from an influx of EV owners, we should also expect more EV owners to use public charging stations because charging a single EV can increase household electricity consumption by 50% [7]. As public-charger demand grows, these stations will become essential and must have availability to build consumer confidence in their greater adoption [17]. An additional 10–20% EV market penetration would increase the daily peak electricity demand by 17.9–35.8% [18]. Such high electricity peaks may cause outages and other issues. High EV penetration and the resulting losses in the network would consequently impose more complexity on the solution of the EV-charger application problem [19]. To solve these issues, calculated maps are needed indicating the optimal city areas where charging stations could be placed according to specific scoring levels (which, of course, depend on the weighting factors). Space limitations and the maximum acceptable distance from the electricity network must also be considered before choosing new EV charger locations [20].

Zhou et al. [6] investigated the relationships between income and affluence levels and the tendency to acquire an EV in Ireland and examined the private EV-household-charger population's characteristics using a regression model and spatial analysis. The Ireland study used information on EV household chargers rather than EV ownership due to data limitations. "The results indicate that 1) urban areas are more likely to see higher concentrations of EV ownership, 2) an income and equity gap exists .... This finding is very important because it suggests that lower-income categories may have a financial barrier to shifting to EVs" [21].

It is very common to use ArcGIS tools to identify suitable locations to install EV charging stations. For example, [22] suggested installing DC fast-charging stations at public libraries and parks within 0.5 miles of major freeways based on the GIS data [23], and Chen suggested setting EV charging stations close to McDonald's and

Starbucks using geospatial datasets [24]. While the development of EV charging stations is in progress, several studies have noted that the accessibility of charging stations is one of the barriers to the adoption of EVs, especially for people with low incomes and those who live in multifamily housing [25]. The distribution of EV charging stations was uneven [26]. In some cities, the location of EV charging stations was not determined by population density, but by people's income [27], and it has caused the issue of inequitable access to EV charging stations. When the government is involved in developing EV charging stations, it becomes essential to "distribute the benefits of facilities to all stakeholders" [28].

Our underlying theory is that EV charging stations are unevenly distributed throughout the city of Los Angeles. It is commonly seen throughout past research that affluent areas tend to be better equipped with EV infrastructure. Although previous analysis assessed EV-charging-station placement, we took a distinct approach because our research analyzes several factors, including current EV stations, area median income, EVs per Zip Code, and population. Previous research only analyzes one or two of these factors. We decided to combine these factors in addition to calculating a current need based on the EV-to-charger ratio, allowing us a better understanding of where and how many EV chargers need to be placed moving forward.

Our fundamental approach is to analyze the correlation between EV-charger distribution, area income, and EV ownership in each area. This will help determine where and how many EV charging stations are needed to meet the current administration's goal.

III. PROBLEM DEFINITION

California has the most EVs in the country, 40% of all EVs in the United States. The problem is our current EV infrastructure is unevenly distributed and will fail to meet the needs and demands of future EV drivers. Since the recent legislature, [1] and [2] have been enacted, we expect an influx of EV drivers, and therefore, more EV chargers must be readily available to meet the public's demand.

Currently, Los Angeles County still has areas with minimal to no EV-charging availability, although the area has EV owners. To promote EV use and combat limited EV-charger availability, we must determine what areas need EV chargers based on income, population, and demand. The goal is to identify areas where the population owns EVs and has insufficient or no EV chargers. With our analysis, we set out to learn how many and where additional EV charging stations are needed based on the ratio of EV ownership to EV charging station. We investigate how and if EV charger placement is disproportionately affected by income, population, or current EV ownership. Finally, we calculate an EV-to-charger ratio for each Los Angeles Zip Code to identify specific needs, aiming for a goal of 18.7 EVs per charging station.

IV. DATA SELECTION AND ACQUISITION

For our analysis, we used data from the National Renewable Energy Laboratory's developer network to identify alternative fuel stations in the city of Los Angeles. We also used data from the California Department of Motor Vehicles that identifies vehicle ownership by Zip Code and fuel type [29]. Two datasets: *City Boundaries* and *Zip Codes*, came from Los Angeles GeoHub [30] [31]. Lastly, we used area median-income shapefile data from Los Angeles Mayor Eric Garcetti's GeoHub site and total estimates from the Los Angeles Almanac.

Alternative fuel-station data, vehicle ownership data, and Los Angeles Almanac data were downloaded as csv files. *City Boundaries* and *Zip Codes* were imported to ArcGIS as shapefiles. A shapefile was initially used for area median income. A second source of median-income data was located and manually converted to a csv file.

To begin our analysis, we had to solve problems with our data sets. First, our vehicle-ownership file did not contain unique identifiers, so we added these manually before importing the data into ArcGIS. Second, the same data set only contained a Zip Code as a location identifier rather than latitude and longitude; therefore, to map the vehicle ownership detail within each Zip Code on the map, we ran the ArcGIS geocode function to assign coordinates to each Zip Code. In order to perform our OLS analysis, we needed a summary of EV data, a summary of EV-charger data, and area median-income data in one file. Since our initial area median-income data was only a shapefile, we located another source of detailed income data by Zip Code and manually created a csv table containing all three elements.

V. SYSTEM AND METHODOLOGY

We chose ArcGIS Pro, version 3.0, to perform our analysis because it provides intuitive tools that would aid in visualizing and manipulating our data to reveal patterns or correlations with EV-charger placement and other factors.

To evaluate the current distribution of EV charging stations in Los Angeles County, we created a map with three layers: EV charging stations, city boundaries, and Zip Codes. Density analysis shows the concentration or clustering of points or lines on a map. To get a whole picture of the distribution of EV charging stations in Los Angeles County, we used Kernel Density. This tool "calculates a magnitude-per-unit area from point or polyline features using a kernel function to fit a smoothly tapered surface to each point or polyline" [23]. We performed kernel density analysis on the data of EV charging stations in Los Angeles County and evaluated how these charging stations were distributed in the whole county area.

In addition to kernel density analysis, mean center, median center, and directional distribution (standard deviational ellipse) are common useful spatial statistics methods to measure geographic distributions. "Standard deviational ellipse has long served as a versatile GIS tool for delineating the geographic distribution of concerned features" [32]. Kemtec and Knez used standard deviational ellipse with other spatial statistics tools, such as mean and median center, to evaluate the location of EV charging stations in Slovenia [26]. We used mean center to identify the geographic center of EV charging stations in Los Angeles County. Median center helped us locate "the point that minimizes Euclidean distance" to all EV charging stations in Los Angeles County [33]. Directional distribution (standard deviational ellipse) helped us find the central tendency, dispersion, and directional trends of EV charging stations in Los Angeles County.

Summarize Within, a statistics tool in ArcGIS, very effectively allows users to "overlay a polygon layer with another layer to summarize [and] calculate attribute field statistics about the features within the polygons" [23]. We used this tool to investigate the locations of EV charging stations in each city and each area by Zip Code. For insight into the number of EV charging stations in each city, we set the city boundaries as Input Polygons and set EV charging stations as Input Summary Features, and then calculated the total EV charging stations in each city. Similarly, we calculated the total EV charging stations in each area based on Zip Codes.

We then calculated EV-to-charger ratios within each Zip Code using Summarize Within within Zip Code boundary polygons. We partially joined the summarized tables containing the EV-ownership detail and EV-charger detail and added a field that divided the sum of EV ownership by the sum of EV charging stations. For the purpose of identifying critical null values in Zip Codes where EV ownership is high and there are no EV charging stations, we manually assigned a −1 value to distinguish these areas in our visualization. We added another field to calculate the number of EV stations needed to bring the currently calculated ratio up or down to 18.7. With this newly created table, we performed a Hot Spot analysis to identify concentrations of high and low EV charger placement. Additionally, we used Optimized Outlier Analysis to identify areas of high and low EV charger placement, as well as to identify areas of high and low EV charger concentration. Finally, we created a manual table containing summarized data per Zip Code for EV ownership, EV-charger placement, and area median income. With this, we use OLS to find correlations between EV-charger placement and area median income or EV ownership.

VI. RESULTS/DISCUSSION

Figure 1 presents a map of Los Angeles County that displays EV charging stations and population density. The map was created to show disparities between areas with high and low population access to EV chargers.

Figure 2 presents a close-up group showing high-population to high-EV-charger density in Los Angeles County based on Figure 1. Areas with "high population density" are areas with a population of 22,000 or more and are shown as orange to yellow shades on the map.

Figure 3 is a close-up group that displays low-population to low-EV-charger density in Los Angeles County based on Figure 1. Areas with "low population density" are areas with a population of 22,000 or less and are shown as purple to black shades on the map.

These two figures show disparities between areas with high population access to EV chargers and low population access to EV chargers. The average distance from EV charger to EV charger in an area with high population density is 0.341 miles. The same average distance in areas with low population density is 0.913 miles. This shows an accessibility disparity, proving that areas with higher populations have more access to EV chargers. This is an issue because high population density does not necessarily point to EV-charger use.

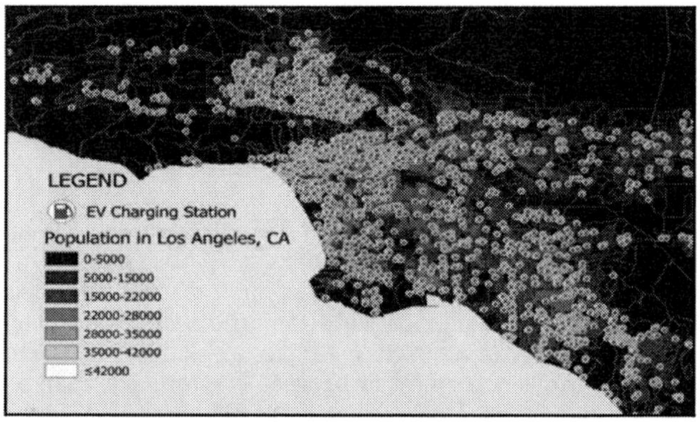

Figure 1. Los Angeles County EV-to-charger ratios and populations.

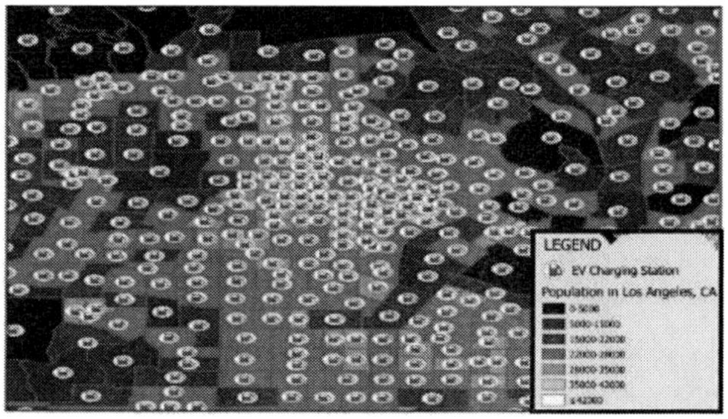

Figure 2. High-population to high-EV-charger density in Los Angeles.

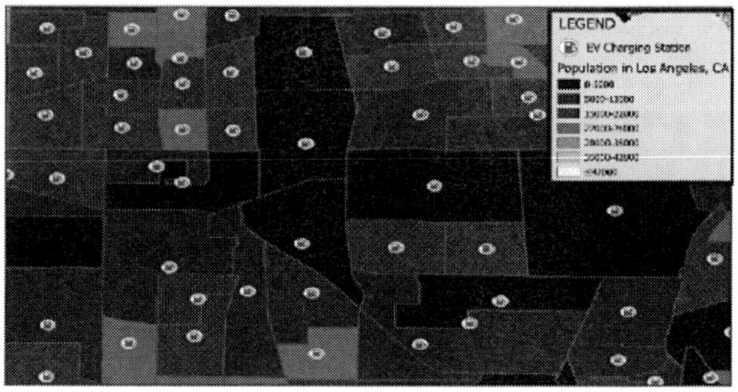

Figure 3. Low-population to low-EV-charger density in Los Angeles.

The kernel density of EV charging stations in Los Angeles County (Figure 4) shows the highest densities of EV charging stations are in Downtown Los Angeles; others are in Pasadena, Hollywood, Beverly Hills, Santa Monica, Los Angeles International Airport, and Long Beach. The western regions of Los Angeles County have more charging stations than the eastern regions.

Figure 5 shows that the mean and median centers of EV charging stations are located near Downtown Los Angeles, and the standard deviational ellipse shows the first standard deviation area: Around 68% of EV charging stations are in this ellipse.

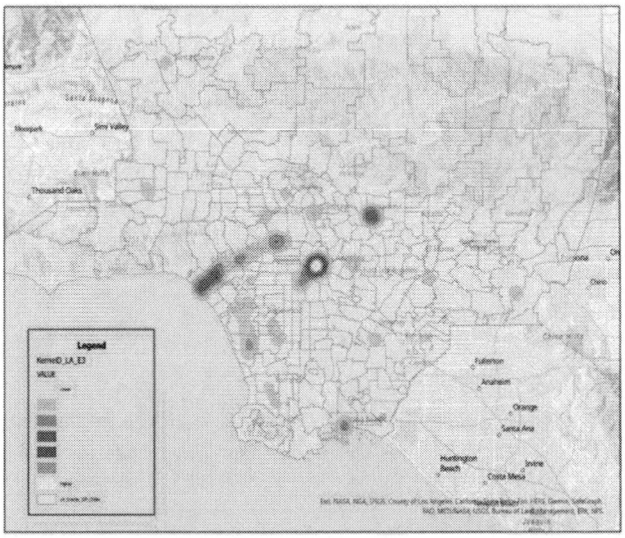

Figure 4. Kernel density of EV charging stations in Los Angeles County.

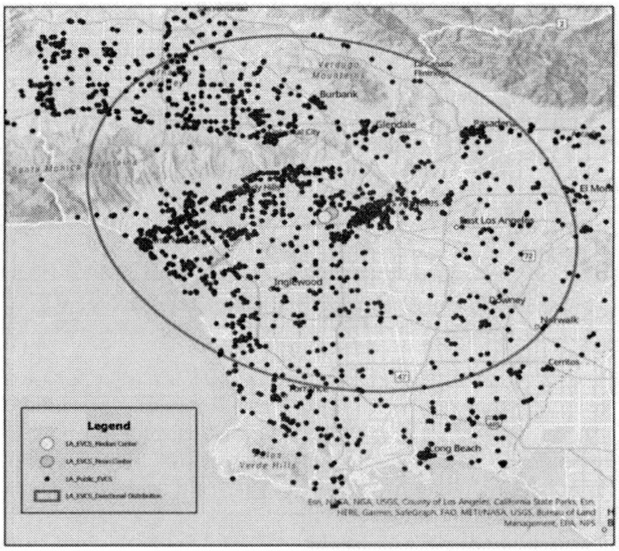

Figure 5. Mean center, median center, and directional distribution of EV charging stations in Los Angeles County.

As the distribution of EV charging stations in each city shows (Figure 6), most EV charging stations are in the city of Los Angeles, the red area. The bar chart (Figure 7) shows a total of 3,028 chargers there, which accounts for 34% of the county's total EV charging stations.

The distribution of EV charging stations in the area by Zip Code (Figure 8) shows how they are located in each Zip Code. The area in red (light yellow) indicates more (fewer) charging stations. Table I lists the top 10 areas by the number of charging stations, and it clearly shows that most of these areas are from the top four cities in which most EV charging stations are located, Los Angeles, Santa Monica, Long Beach, and Pasadena.

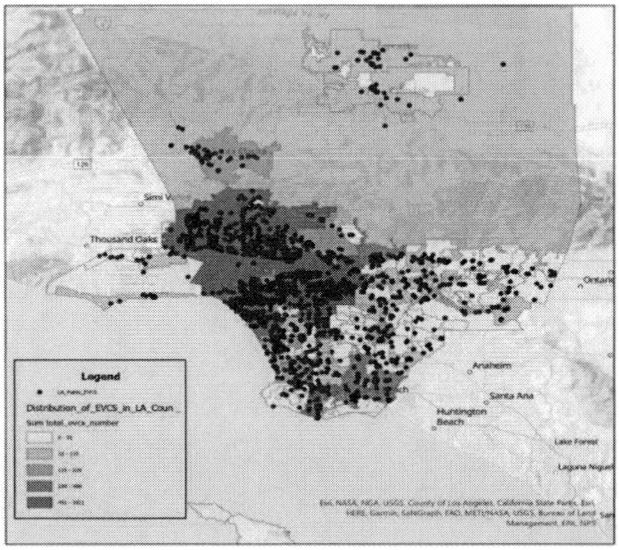

Figure 6. Distribution of EV charging stations in each city.

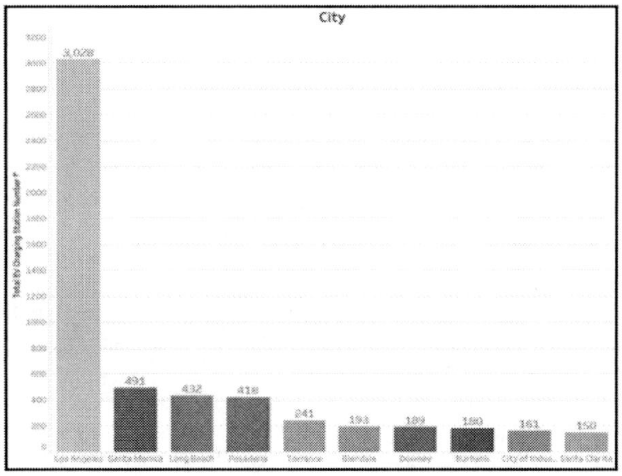

Figure 7. Top 10 cities by the number of total EV charging stations.

TABLE I. TOP 10 ZIP CODE AREAS BY THE NUMBER OF EV CHARGERS

	Zip Code	City	Number of EV Charging Stations
1	90012	Los Angeles	351
2	90045	Los Angeles	200
3	90802	Long Beach	193
4	90404	Santa Monica	191
5	91355	Santa Clarita	185
6	90028	Los Angeles	175
7	90007	Los Angeles	162
8	90401	Santa Monica	143
9	90245	El Segundo	143
10	91125	Pasadena	139

Our analysis found an uneven distribution of EV charging stations in Los Angeles County. Most charging stations are located in the western region of the county, concentrated in Downtown Los Angeles. This finding suggests that the accessibility of EV charging stations varies by area. People living around Downtown Los Angeles have more convenient access to charging stations.

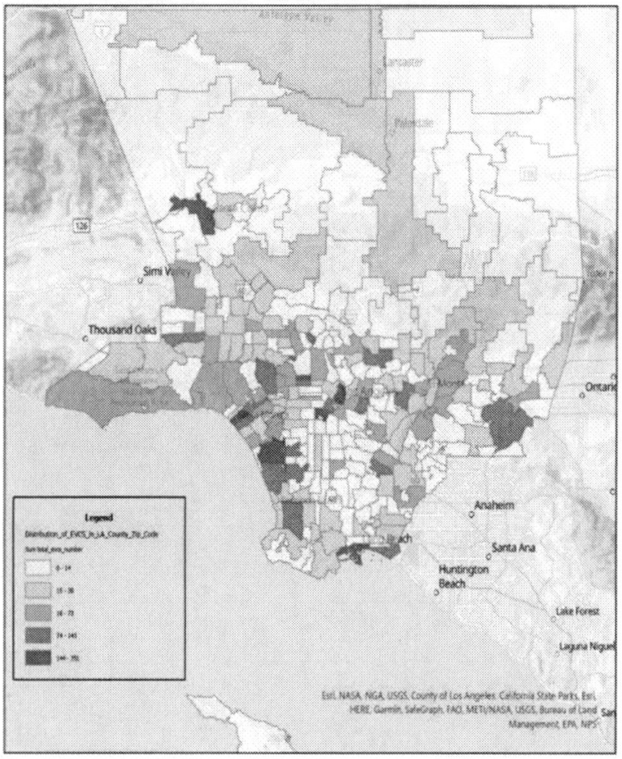

Figure 8. Distribution of EV Charging Stations in the Area by Zip Code.

OLS of the manually created table contains the sum of EV charging stations as a dependent variable and the sum of EV ownership and area median income as explanatory variables. Results R^2 multiple of 8% explained no correlation among these variables in Los Angeles County (Figure 9).

To complete the following analysis, we calculated the EV-to-charger ratio by dividing the sum of EV ownership within each Zip Code by the sum of EV charging stations to determine a ratio within each Zip Code (Table II). By doing so, we will identify specific Zip Codes where the ratio is null, indicating EVs but no charging stations, or Zip Codes where the EV-to-charger ratio is high and in need of additional stations to bring the ratio down to the current administrations' goal of 18.7.

TABLE II. SAMPLE DATA OF LOS ANGELES COUNTY EV OWNERSHIP TO EV CHARGING STATION RATIO CALCULATIONS

Zip Code	Sum of Electric Vehicles	Sum of EV charging stations	EV to charging station Ratio
90059	36	5	7.2
91301	180	5	36.0
91775	*335*	*0*	*−1**
91801	345	9	38.3
90007	68	65	1.0

*Indicates placeholder number not actual calculation

A hot/cold spot analysis of EV charger locations that only includes areas where the ratio is greater than 18.7 as proposed by the current administration (Figure 10) shows a higher concentration of EV chargers in northwest Zip Codes from Los Angeles city center and low concentrations in Zip Code South East of the Los Angeles city center.

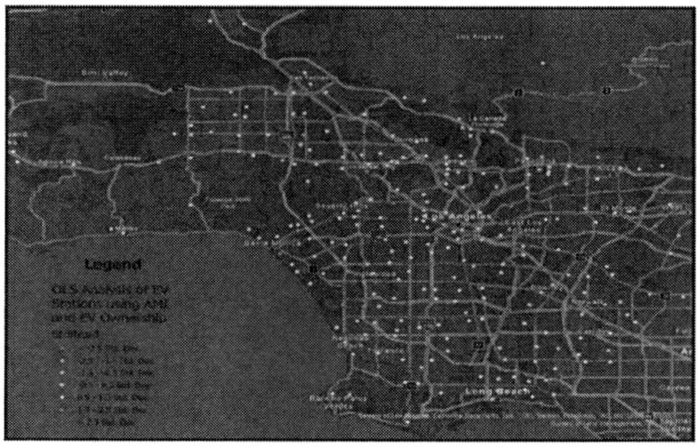

Figure 9. OLS analysis of Los Angeles EV charging station placement using AMI and vehicle ownership as explanatory variables.

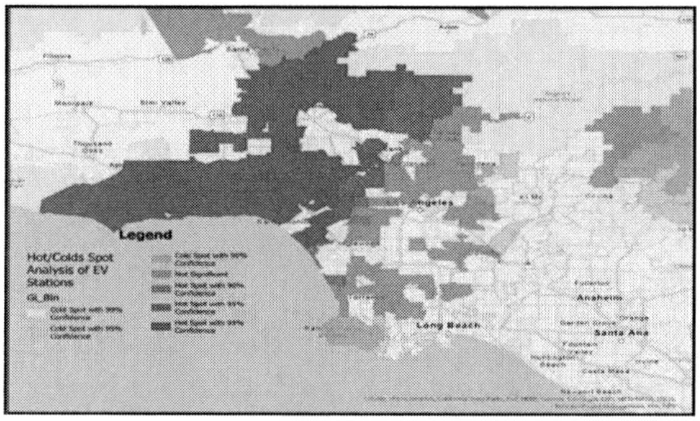

Figure 10. Hot/cold spot analysis of EV chargers within areas of Los Angeles with calculated EV-to-charger ratio greater than 18.7.

An optimized-outlier analysis of EV charger locations that only includes areas where the EV-to-charger ratio is greater than 18.7 as proposed by the current presidential administration (Figure 11) highlights areas of urgent need for EV charging stations in light blue and red. Light blue highlights output features with clusters of low sums of EV chargers, and red highlights output features with high outliers within a cluster of low sums of EV-charger values.

A visualization of the EV-to-charger ratio in all areas broken out by Zip Code (Figure 12) indicates that the majority of Los Angeles Zip Codes falls significantly far above the proposed ratio of 18.7. This means there are more than 18.7 EVs per charger in these Zip Code, which could lead to availability issues.

Upon identifying areas in need of more EV chargers, we identified how many EV charges are needed within each Zip Code in areas where the EV-to-charger ratio is greater than 18.7 (Figure 13).

Figure 14 presents an identified area in need of more EV chargers to display the specific locations of existing EV chargers. Our calculation shows that these sample areas in

Los Angeles require between 19 to 24 new EV chargers to bring the current EV-to-charger ratio down to 18.7 and ensure availability.

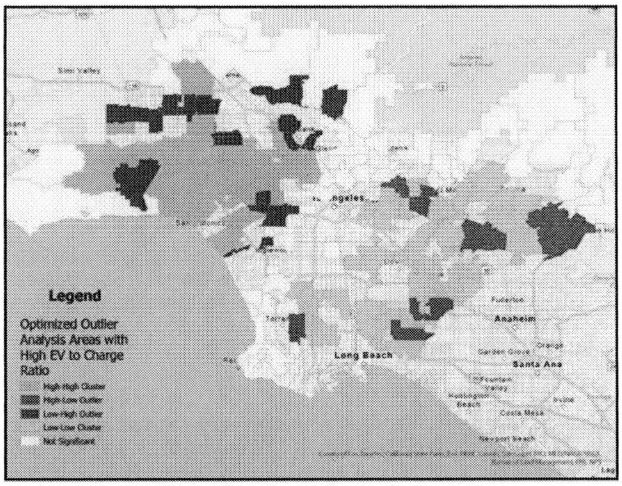

Figure 11. Optimized outlier analysis of EV chargers in areas of Los Angeles with calculated EV-to-charger ratios greater than 18.7.

Figure 12. Los Angeles County EV-to-charger ratio all areas.

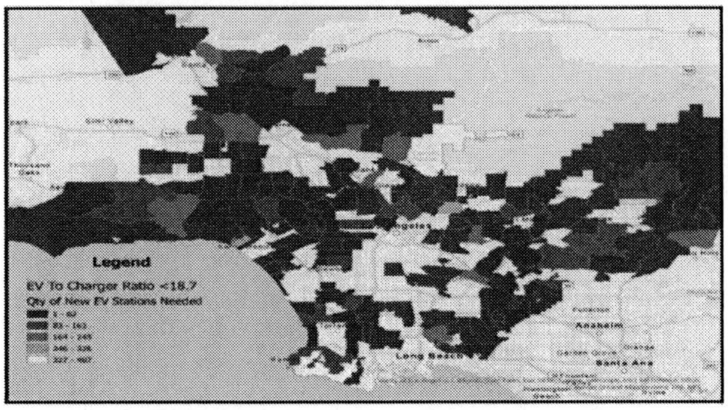

Figure 13. Quantity of new EV charging stations needed within areas of calculated ratio greater than 18.7 in Los Angeles County.

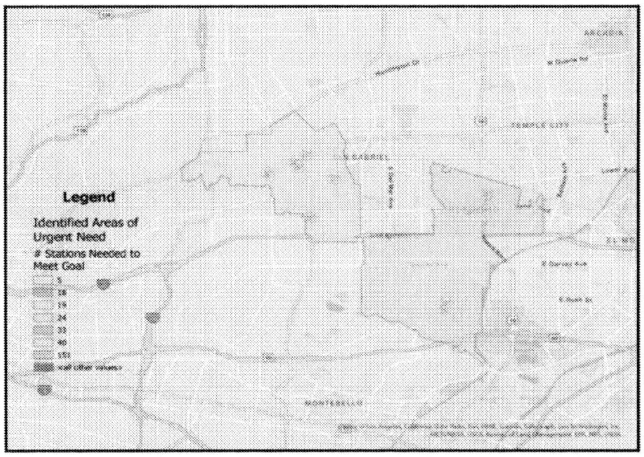

Figure 14. Sample area in Los Angeles County with identified quantity of new EV charging stations needed to meet goal ratio of 18.7.

VII. Project Limitations

With access to more detailed spatial data, our analysis could be more in depth. For example, access to privately owned EV charger providers, such as Tesla, EVgo, and ChargePoint, would allow us to analyze their trend, peak time, charge frequency, and public reviews to correlate public preferences and determine the need for higher quality and more reliable charger placements. Using a dataset within ArcGIS pro that does not contain latitude and longitude limits the analysis. Since vehicle ownership data is personal and private, our dataset was limited to quantities per Zip Code and therefore the rest of our analysis was modeled around Zip Code analysis. Exact location data would have provided exact walking distances and charger accessibility.

VIII. Conclusion

Our analysis was able to determine by visual observation that areas in Los Angeles with high populations tend to have more EV chargers. Kernel-density, spatial, hot-spot, and outlier analysis confirmed that the distribution of EV chargers in Los Angeles County is uneven: Areas north and west of Los Angeles city center have higher concentrations of EV chargers, while areas east and south of the city center have significantly lower concentrations. Most EV chargers are located in the city of Los Angeles, especially in Downtown Los Angeles. Such an uneven distribution of EV chargers would cause inequitable access issues for people living and working in other areas.

Upon categorizing each Zip Code by EV-to-charger ratio, we specifically identified areas where additional EV chargers are needed and, more importantly, determined how many EV chargers would be required within each Zip Code. For Los Angeles to become fully electric by 2035, investment in EV infrastructure will require the implementation of many new EV chargers to support current and future EV drivers. Ensuring that the EV-to-charger ratio is lowered and remains low will ensure that drivers are able to access EV chargers when they need them.

Some possible extensions of this work are modeling EV penetration in different population areas and examining the distribution of vehicle-brand-neutral charging stations. Also, studies could examine the effects of encouraging low-income drivers to use EVs: Upfront costs for the drivers, the ecological impact of recycling batteries and mining the resources to make more batteries. Finally, more private data should be available at all granularities to improve on studies like this.

REFERENCES

[1] State of California. *Governor Brown takes action to increase zero-emission vehicles, fund New Climate Investments.* Governor Edmund G Brown Jr. [Online]. Available from: https://www.ca.gov/archive/gov39/2018/01/26/governor-brown-takes-action-to-increase-zero-emission-vehicles-fund-new-climate-investments/index.html

[2] The White House. *Fact sheet: The Biden–Harris electric vehicle charging action plan.* [Online]. Available from: https://www.whitehouse.gov/briefing-room/statements-releases/2021/12/13/fact-sheet-the-biden-harris-electric-vehicle-charging-action-plan/

[3] A. Bui, P. Slowik, and N. Lutsey. *Los Angeles electric vehicle charging infrastructure needs and implications for zero-emission area planning.* [Online]. Available from: https://theicct.org/publication/los-angeles-electric-vehicle-charging-infrastructure-needs-and-implications-for-zero-emission-area-planning/

[4] P. Chakraborty et al., "Addressing the range anxiety of battery electric vehicles with charging en route," Sci. Rep., vol. 12, no. 1, pp. 1–15, 2022.

[5] D. Mayfield. *Siting electric vehicle charging stations.* Ed. Carlotta Collette. Sustainable Transportation Strategies. [Online]. Available from: https://vacleancities.org/wp-content/uploads/Siting-EV-Charging-Stations-FINAL-12.pdf

[6] G. Zhou, Z. Zhu, and S. Luo, "Location optimization of electric vehicle charging stations: Based on cost model and genetic algorithm," Energy, vol. 247, art. 123437, 2022.

[7] M. B. Arias and S. Bae, "Electric vehicle charging demand forecasting model based on big data technologies," Appl. Energy, vol. 183, pp. 327–339, 2016.

[8] A. G. Abo-Khalil et al., "Electric vehicle impact on energy industry, policy, technical barriers, and power systems," Int. J. Thermofluids, vol. 13, art. 100134, 2022.

[9] H. Chih-Wei and K. Fingerman, "Public electric vehicle charger access disparities across race and income," Transp. Policy, vol. 100, pp. 59–67, 2021, doi:10.1016/j.tranpol.2020.10.003

[10] A. Roy and M. Law, "Examining spatial disparities in electric vehicle charging station placements using machine learning," Sustain. Cities Soc., vol. 83, art. 103978, 2022.

[11] Y. He, K. M. Kockelman, and K. A. Perrine, "Optimal locations of US fast charging stations for long-distance trip completion by battery electric vehicles," J. Clean. Prod., vol. 214, pp. 452–461, 2019.

[12] H. Lin, et al., "Optimal planning of intra-city public charging stations," Energy, vol. 238, art. 121948, 2022.

[13] M. Gellrich, A. Block, and N. Leikert-Böhm, "Spatial and temporal patterns of electric vehicle charging station utilization: A nationwide case study of Switzerland," ERIS, vol. 2, no. 2, art. 021003, 2022. doi:10.1088/2634-4505/ac6a09

[14] Y. Zhou, D. Gohlke, M. Sansone, J. Kuiper, and M. P. Smith. *Using mapping tools to prioritize electric vehicle charger benefits to underserved communities.* No. ANL/ESD-22/10. Argonne National Lab. Argonne, IL (United States).

[15] M. Nazari-Heris, A. Loni, S. Asadi, and B. Mohammadi-Ivatloo, "Toward social equity access and mobile charging stations for electric vehicles: A case study in Los Angeles," Appl. Energy, vol. 311, art. 118704, 2022.

[16] S. LaMonaca and L. Ryan, "The state of play in electric vehicle charging services—A review of infrastructure provision, players, and policies," Renewable Sustainable Energy Rev., vol. 154, art. 111733, 2022.

[17] F. Avalos Jr, "An evaluation of electric vehicle charging stations with respect to functional consumer demand," Doctoral dissertation, University of California, Los Angeles.

[18] M. A. Quddus, O. Shahvari, M. Marufuzzaman, and S. D. Eksioglu, "Designing a reliable electric vehicle charging station expansion under uncertainty," Int. J. Prod. Econ., vol. 236, art. 108132, 2021. doi:10.1016/j.ijpe.2021.108132

[19] H. Simorgh, H. Doagou-Mojarrad, H. Razmi, and G. B. Gharehpetian, "Cost-based optimal siting and sizing of electric vehicle charging stations considering demand response programmes," IET Gener. Transm. Distrib., vol. 12, no. 8, pp. 1712–1720, 2018.

[20] D. Gkatzoflias et al., *Optimal allocation of electric vehicle charging infrastructure in cities and regions.* The new technical research report of the Joint Research Centre of the European Commission, the European Commission's in-house science service. 2016.

[21] B. Caulfield, D. Furszyfer, A. Stefaniec, and A. Foley, "Measuring the equity impacts of government subsidies for electric vehicles," Energy, vol. 248, art. 123588, 2022.

[22] J. S. Jin, "Installing public electric vehicle charging stations: A site suitability analysis in Los Angeles County, California," Doctoral dissertation, University of Southern California.

[23] ArcMap. *An overview of the Density toolset.* [Online]. Available from: https://desktop.arcgis.com/en/arcmap/latest/tools/spatial-analyst-toolbox/an-overview-of-the-density-tools.htm

[24] J. Y. L. Chen, "California public electric vehicle charging stations' accessibility to amenities: A GIS network analysis approach." Master's Projects and Capstones. [Online]. Available from: https://repository.usfca.edu/capstone/565

[25] S. Hardman, K. Fleming, E. Khare, and M. M. Ramadan, "A perspective on equity in the transition to electric vehicles," MIT Sci. Pol. Rev., vol. 2, pp. 46–54, 2021. doi:10.1088/2634-4505/ac6a09

[26] M. Kmetec and M. Knez, "Electric vehicle charging stations coverage: A study of Slovenia," Tehnički Vjesnik, vol. 29, no. 1, pp. 285–292, 2022.

[27] H. A. Khan, S. Price, C. Avraam, and Y. Dvorkin, "Inequitable access to EV charging infrastructure," Electricity Journal, vol. 35, no. 3, art. 107096, 2022. doi:10.1016/j.tej.2022.107096

[28] B. D. Chung, S. Park, and C. Kwon, "Equitable distribution of re-charging stations for electric vehicles," Socio-Econ. Plan. Sci., vol. 63, pp. 1–11, 2018, doi:10.1016/j.seps.2017.06.002

[29] State of California. *Vehicle fuel type count by ZIP code 1/1/2021.* [Online]. Available from: https://data.ca.gov/dataset/vehicle-fuel-type-count-by-zip-code/resource/888bbb6c-09b4-469c-82e6-1b2a47439736

[30] Los Angeles GeoHub. *City boundaries. City of Los Angeles Hub.* [Online]. Available from: https://geohub.lacity.org/datasets/lacounty::city-boundaries-3/explore?location=33.886095%2C-118.390855%2C8.28

[31] Los Angeles GeoHub. *ZIP codes (La County).* [Online]. Available from: https://geohub.lacity.org/datasets/lahub::zip-codes-la-county/explore?location=33.795647%2C-118.302795%2C9.00

[32] B. Wang, W. Shi, and Z. Miao, "Confidence analysis of standard deviational ellipse and its extension into higher dimensional Euclidean space," PloS One, vol. 10, no. 3, pp. e0118537–e0118537, 2015. doi:10.1371/journal.pone.0118537

[33] Los Angeles GeoHub. *Median income and AMI (census tract).* [Online]. Available from: https://geohub.lacity.org/datasets/lacounty::median-income-and-ami-census-tract/explore?location=33.769264%2C-118.302668%2C9.00

Business Model Considerations for a Solution to Optimize and Diagnose Solar Panel Installations

Lasse Berntzen
University of South-Eastern Norway
Horten, Norway
email: lasse.berntzen@usn.no

Saeed Teimourzadeh
EPRA Electric Energy Co.
Ankara, Turkey
email: saeed@epra.com.tr

Paula Anghelita
ICPE SA
Bucharest, Romania
email: p.anghelita@icpe.ro

Qian Meng
University of South-Eastern Norway
Horten, Norway
email: qian.meng@usn.no

Viorel Ursu
ICPE SA
Bucharest, Romania
email: vio.ursu@icpe.ro

***Abstract* — This paper describes the business model considerations for an Internet of Things (IoT) solution to diagnose and optimize solar panel installations. The solution consists of a sensor platform and a cloud-based service. The sensor platform collects information about the position and orientation of the solar panels, the solar radiation, the ambient temperature and humidity, and the temperature of the solar panels. The cloud-based service receives the collected data together with data from the inverter. An algorithm analyzes the data and makes recommendations about possible adjustments and malfunctions. Users are offered recommendations through a web-based interface. The focus of this paper is the business model considerations. How can the solution create value for the manufacturer and stakeholders in the solar panel business? The solar panel value chain is examined to see where the solution can be bundled with other business activities.**

Keywords-solar panels; photovoltaic energy; testing; business model; IoT; cloud-based service.

I. INTRODUCTION

Prosumers are energy consumers that also produce energy. The produced energy can be used, stored for later consumption, or sold to the grid. A smart meter keeps track of the energy flow between the prosumer and the grid. Fig. 1 shows a prosumer connected to the grid through a smart meter. The prosumers have solar panels on their rooftops to produce their own energy.

The most popular technology for small-scale renewable production is solar panels. However, solar panels are only effective during the daytime, with a peak output when the solar radiation is the highest. A solar panel needs to capture as much radiation as possible. Therefore, the characteristics and the orientation/positioning of the solar panel are critical factors for the efficiency of energy production. The output of the solar panels also depends upon the surface temperature of the solar panel. High temperature dramatically decreases photovoltaic output.

To face this challenge, a collaborative effort of teams from three countries with different climatic conditions, ICPE SA from Romania, EPRA Electric Co. from Turkey, and the University of South-Eastern Norway joined their efforts in the project *Cloud-based analysis and diagnosis platform* for *photovoltaic (PV) prosumers* to develop a solution to optimize and diagnose solar panel installations. The solution was named Dr. Solar since it diagnoses the solar panel installation. The solution, including the algorithms used for optimization, is described in detail in [1]. Section 3 provides an overview of the solution and the sensor platform used.

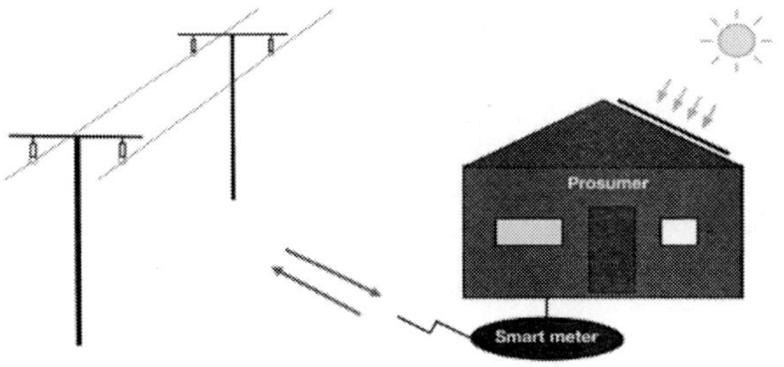

Figure 1. The prosumer

This paper aims to discuss possible business models for the Dr. Solar solution. Section 2 briefly introduces the Internet of Things. Section 3 describes the Dr. Solar solution. Section 3 discusses business models in general, while Section 4 considers business models for the Dr. Solar solution. Section 6 presents the results, while Section 7 provides a conclusion and ideas for future work.

II. INTERNET OF THINGS

The Internet of Things (IoT) refers to devices connected to the Internet. The connection to the Internet opens new opportunities. IoT devices can collect information from sensors and control the environment through actuators. The network makes it possible to process information at a remote location. Fig. 2 shows a couple of IoT devices connected to an IoT service through the Internet.

A cloud-based service processing data reduces the need for local processing capacity in the IoT device. The cloud-based service also makes it easier to replace or tune the algorithms used.

.

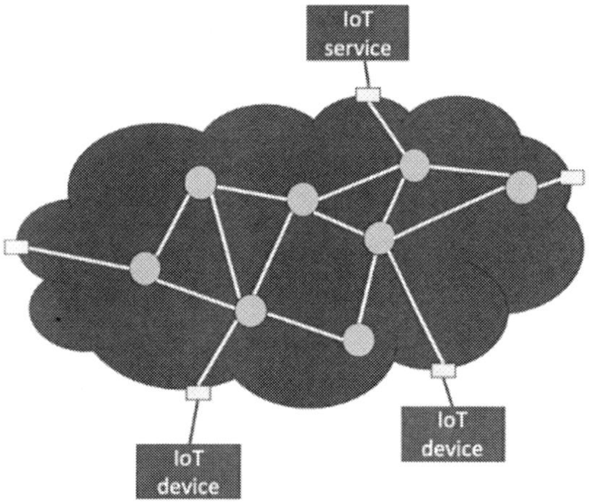

Figure 2. Internet of Things

One of the typical IoT applications in the energy sector is monitoring power output and related parameters, such as voltage, current, and consumption. Such monitoring can support short- and long-term planning of operations as well as the future development of the power grid. Monitoring may require real-time or close to real-time data transmissions. Real-time requirements may influence the design of the IoT architecture. The value of IoT lies in the data the sensors provide and how the data is interpreted. A cloud-based service can analyze sensor data and provide decision-making information

III. THE DR. SOLAR SOLUTION

This section describes the solution and the sensor platform.

A. The Solution

Dr. Solar is an IoT solution developed to assess the performance and operation of photovoltaic systems. The solution consists of three main subsystems, shown in Fig. 3.

The first subsystem is the Dr. Solar sensor platform containing sensors and data loggers. The Dr. Solar sensor platform acts as the data acquisition system for the Dr. Solar solution. The sensor platform collects and uploads data to the cloud-based service using a GSM-based mobile router. A router is necessary since the solar panels are usually not placed within the range of an existing Wi-Fi network. The sensor platform is configured to collect data every 5 minutes for three consecutive days, ideally when solar radiation exceeds 600 W/m2. The sensor platform and the inverter provide data at five-minute intervals polled by the cloud-based analytics service for the time interval between 10:00 am to 2:00 pm. More details on the sensor platform will be discussed later.

The second subsystem is the inverter which converts the generated DC power by the solar panels into AC power. Within this process, the inverter collects and provides data concerning energy production. The prototype of the Dr. Solar solution uses a Huawei inverter SUN2000L-3KTL model to demonstrate the functionality.

The third subsystem is a cloud-based analytic service that contains assessment algorithms and engines. The engines within the cloud-based service acquire inputs from the Dr. Solar sensor platform and the inverter to perform two main tasks:

1. Performance evaluation which is called asset management service in Dr. Solar solution. The main objective of the asset management service is to evaluate the performance of the solar panels and perform a health check. The solution provides information about the problem and possible reasons to the customer if any malfunction is observed.
2. Performance enhancement which is called optimum decision support in the Dr. Solar solution. The optimum decision support service's main objective is to evaluate the solar system's operating conditions and develop an optimal operation schedule to reduce the customer's electricity bill. The optimum decision support considers user preferences and uses these together with the data from the inverter. The optimum decision support does not rely on the Dr. Solar sensor platform and can be done after the unit is removed from the installation.

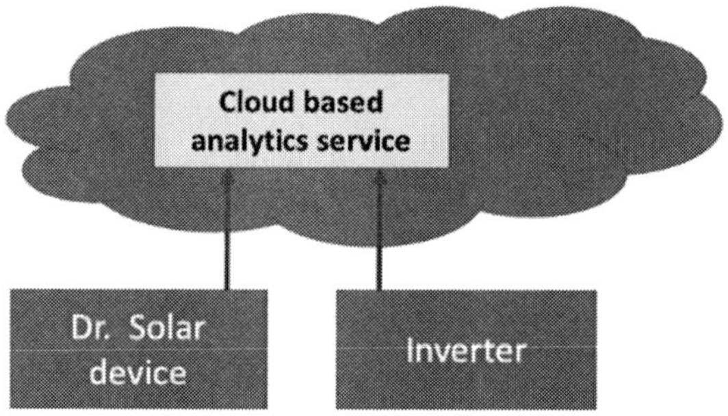

Figure 3. The Dr. Solar solution

The Dr. Solar solution provides a user-friendly web interface to investigate monitoring results for asset management and optimum decision support.

B. The Sensor Platform

The Dr. Solar sensor platform is the hardware component of the solution, including sensors for data acquisition and submission to a dedicated cloud-based analysis and diagnosis platform. The sensor platform is compact, autonomous, and collects data regarding:

- Direct solar radiation level in the plane of the photovoltaic generator (solar panel).

- Ambient temperature and humidity, as well as solar panel temperature.
- Geographical location and specific installation angles (the inclination and the azimuthal deviation of the solar panel from the southern direction).

The sensor platform is easy to mount onto the solar panel installation, and a prosumer can install the unit without expert help.

The architecture of the IoT sensor platform is shown in Fig. 4. The different parts communicate through signals and protocols to collect sensor information.

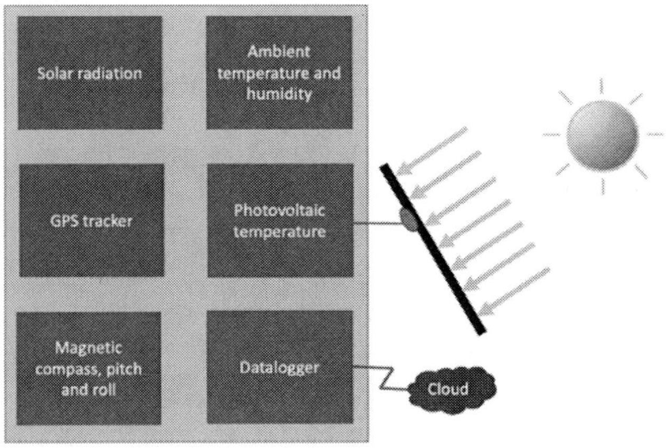

Figure 4. The architecture of the sensor platform

From the sensor platform, the data is transferred through web get-type procedures to the remote subsystem residing in the cloud.

Figure 5. The physical construction of the sensor platform

Fig. 5 shows the physical construction of the sensor platform, while Fig. 6 shows how the sensor platform is mounted on a solar panel.

IV. IoT Business Models

A business model is the underlying knowledge of a venture or the core logic for generating value in a venture [2][3]. Research has tried to establish a more unified business model by using it as a new analytical unit with a system-level, holistic approach to explain how a company creates value through its activities [4]. A holistic approach includes the role of the market, the customer, and the systematic and integrated approach to business model innovation processes [5]. Considering the relationship between explicit needs and extended value chain, a business model is defined as a mechanism to comprise the framework/mode/logic of profit-seeking and business structure to make a profit [6].

Figure 6. The sensor platform mounted on a solar panel

A business can expand its value creation by increasing its activities upstream, downstream, or both. In our case, the industrial partner, ICPE, delivers solar panel installations. Upstream activities would be producing solar panels, while downstream activities may be operating and maintaining the solar panel installations. The Dr. Solar box represents downstream value creation.

IoT opens new opportunities for business models. Gassmann, Frankenberger, and Csik [7] conceptualized the business model by four central dimensions: the *who*, the *what*, the *how*, and the *value*. Who is the target customer or segment? What do we offer to the customer (value proposition)? How do we deliver our goods and services (value chain)? And finally, how do we generate value (revenue model)? The conceptualization is shown in Fig. 7.

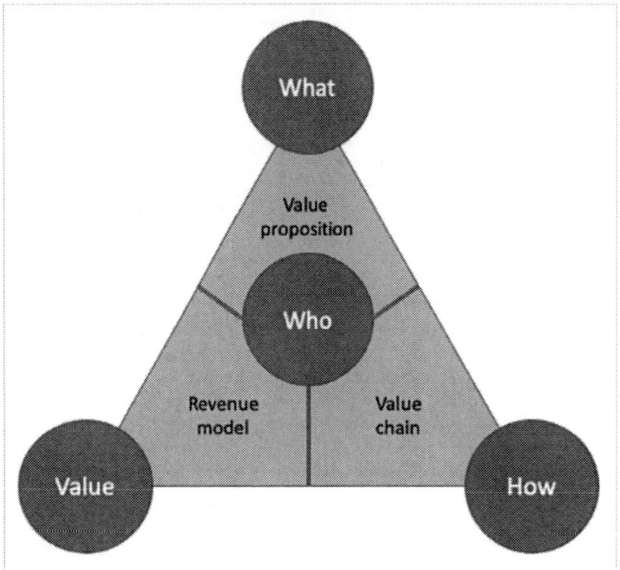

Figure 7. Central dimensions of a business model [7]

Sinclair [8] presents five classes of business models and shows the evolution over time, called the business model continuum. The first class is called the *product* business model, where a product is sold to the customer. IoT can provide added value through new features, better performance, reduced cost, and increased accuracy. The second class is the *product-service* business model. Here, the relationship between the producer and the customer creates added value. The relationship can incorporate training, predictive maintenance, and smart services.

TABLE I. REVENUE MODELS

Business Model Class	Description	Revenue Model
Product	The customer buys the hardware unit and gets access to the service for free.	Revenues come from the sale of the hardware unit.
Product/ service	The customer buys the hardware unit and subscribes to a service.	Revenues come from the sale of the hardware unit and the subscription fee.
Service	The customer rents the hardware unit or subscribes to a service, and the service provider owns the hardware. The model facilitates long-term relationships with the customer.	Revenues come from the rent or subscription fee only.
Loan	The customer loans the hardware unit and uses the service for free. It is used to add value to other sales	Revenues come from other sources, e.g., increased sales of solar panels and associated hardware or advertisements on the web platform.
Freemium	The product is installed for free. Functionality is unlocked by subscription.	Revenues come from subscriptions. However, the freemium model has become popular in many segments, but in our case, the unit cost may be too high to gamble on whether customers are subscribing or not.
Service/ outcome	A partnership is created between the manufacturer and a stakeholder in the solar panel value chain	Revenues come from profit sharing, where the manufacturer receives its part from the profits generated by the partnership.
Outcome	The product is installed as part of a total system of solar panels and the inverter	Revenues come from a close partnership with stakeholders responsible for the whole power generation process.

The third class is the *service* business model, where the customer pays per hour of use, per time used, or volume collected. The producer owns the product. The customer subscribes to a service. The last two classes focus on the outcome. The outcome is about solving the customer's problem. The fourth class is the *service-outcome* business model. In this case, the producer installs the necessary products to solve a problem. The monetization can be a share of the value generated by the producer's systems. The *outcome* business model is when the producer and customer establish a close partnership that solves the problem for the customer.

In addition to the five classes proposed by Sinclar [8], we added *loan* and *freemium*. Lending out the hardware unit makes it possible for customers to increase the value of their installations and incentivizes them to choose the manufacturer offering such a service. Freemium [9] implies that essential services are free, while more advanced functionality requires the customer to pay.

A review of business model literature by Osterwalder, Pigneur, and Tucci [10] showed broad diversity of understandings, usages, and places in the firm. Creating value is an important factor. However, Chesbrough [11] and Spencer [12] discuss open business models where many partners participate in the innovation process. Additionally, some marketing researchers are studying co-creation and co-production, in which the client participates in developing the primary offering.

The approach used by Gassmann, Frankenberger, and Csik [7] will now be combined with the classes of business models offered by Sinclar [8] to discuss viable business models for the Dr. Solar solution.

V. DR. SOLAR BUSINESS MODEL CONSIDERATIONS

The value proposition for the Dr. Solar solution is optimizing and diagnosing the solar panel installation. By using Dr. Solar, the user can improve energy production and be alerted about possible malfunctions in the installation.

The revenue model is about creating value for the different stakeholders. Building on the classes proposed by Sinclar [7], we present seven alternatives for creating revenues. The revenue models are shown in Table I.

As to the selling behavior, there are two options. The Dr. Solar solution can be sold separately or bundled with other products, as shown in Table II.

TABLE II. STAND-ALONE OR BUNDLING

Stand-alone	Dr. Solar is offered to customers with existing solar panels
Bundle	Dr. Solar is bundled with other products, e.g., inverter and/or solar panels.

The installation of solar panels involves different stakeholders, as shown in Fig. 8. These stakeholders may bundle the Dr. Solar solution with their own products or activities.

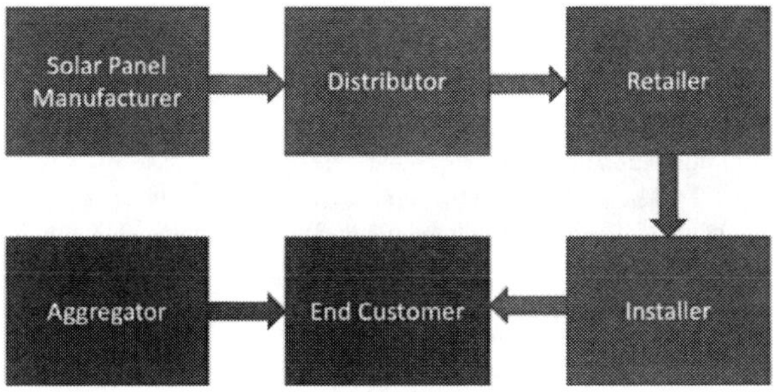

Figure 8. Stakeholders

TABLE III. STAKEHOLDERS

End customer	The end customer would typically use Dr. Solar to diagnose and optimize its photovoltaic installation. The end customer would typically loan the Dr. Solar.
Installer	The installer can use the Dr. Solar unit to improve the installation and achieve a competitive advantage. The installer typically buys and uses the hardware unit for all installation products.
Retailer	The retailer of solar panels and other photovoltaic equipment can offer Dr. Solar on loan or as a service. The retailer could get a competitive advantage from bundling with other products.
Distributor	The distributor could offer Dr. Solar as a product or service to downstream retailers and installers.
Manufacturer	The manufacturer of solar panels and other photovoltaic equipment could offer Dr. Solar as an added value to distributors, retailers, and installers.
Aggregator	The aggregator organizes a group of prosumers. The aggregator could buy Dr. Solar and lend it out to its prosumers.

The stakeholders' use of the solutions is described in Table III, and the bundling opportunities are shown in Fig. 9.

VI. RESULTS

After considering different potential business models, the chosen alternative was to regard Dr. Solar as an added value to photovoltaic installations. The sensors unit, equipped with high performances sensors, is too expensive to be permanently installed on the photovoltaic systems on-site. It is enough to stay installed for only a few days to collect relevant data in one site and then send it to a new installation where the photovoltaic system needs to be assessed. Therefore, the intention is to lend the sensor platform to customers installing photovoltaic systems to diagnose and optimize their installation. Providing this technical solution as a service will obtain a competitive advantage in the market. Bundling with other services may be viable to help create business relationships with other solar panel value chain actors.

The data collected may also be used for value creation. Collecting data from many solar panel installations may be used for analysis on a larger scale. The results can be used to gain further insight into optimizing photovoltaic energy production.

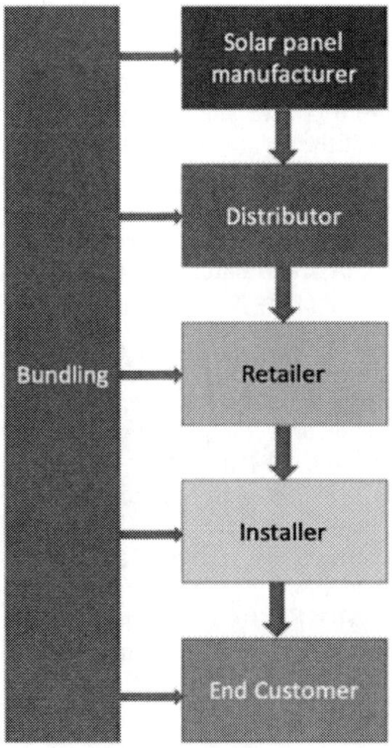

Figure 9. Bundling

VII. CONCLUSION AND FUTURE WORK

The Internet of Things creates new opportunities for innovative business solutions. Dr. Solar is a solution consisting of an IoT sensor platform and a cloud-based service that analyzes sensor data and data from the inverter to optimize and diagnose photovoltaic energy production. This paper focuses on presenting possible business models for deploying the solution in the marketplace. The cost of the sensor platform prohibits permanent mounting to smaller solar panel installations. Therefore, moving the sensor platform from one installation to another is necessary. The solution creates value for the customers by optimizing and diagnosing the installation. However, it is also essential to generate revenues. Several revenue models have been considered. The solar panel installation ecosystem consists of many actors. Bundling is one way of generating revenues.

Dr. Solar is not the final solution; further developments can be done. The application of the solution developed within the PVADIP-C project can be customized to answer further different needs of photovoltaic systems users.

The project did not try to minimize the costs. The components used are all considered high quality. The sensor platform price can be reduced by selecting cheaper components. Selecting other materials for the casing could also bring down the cost.

With lower sensor platform costs, a permanent Dr. Solar unit may be considered for larger photovoltaic installations.

The project has only considered solar panels with a fixed position. The solution provides recommendations on optimal positioning and orientation of the solar panels. However, dynamic positioning can increase energy production. Dynamic positioning is relatively expensive, but enhancing the solution to support the solar panels' real-time positioning would be straightforward. The sensor platform must then be a permanent part of the solar panel installation.

Some of the ideas from Dr. Solar could be used for other types of renewable energy. The sensors would differ, but the overall architecture with a sensor platform and a cloud-based service for diagnosis and optimization could also be used for wind or geothermal energy.

ACKNOWLEDGMENT

This work was part of the project "Cloud-based analysis and diagnosis platform for photovoltaic (PV) prosumers." supported by the Manu Net scheme (Grant no. MNET20/NMCS-3779) and funded through the Research Council of Norway (Grant no. 322500), UEFISCDI - Executive Agency for Higher Education, Research, Development and Innovation Fund (Contract no. 215/2020), and TÜBİTAK ARDEB 1071 Program - Support Program for Increasing Capacity to Benefit from International Research Funds and Participation in International R&D Cooperation (Project no. 120N838)

REFERENCES

[1] L. Berntzen, S. Teimourzadeh, P. Anghelita, Q. Meng, and V. Ursu, "A Sensor-based Unit for Diagnostics and Optimization of Solar Panel Installations," 19th European Mediterranean & Middle Eastern Conference on Information Systems. Lecture Notes in Business Information Processing. Springer, 2023 (To be published)

[2] S. M. Shafer, H. J. Smith, and J. C. Linder, "The Power of Business Models," Business Horizons, vol 48(3), 2005, pp. 199-207.

[3] M. Malmstöm, "Typologies of Bootstrap Financing in Small Ventures," Venture Capital: An International Journal of Entrepreneurial Finance, vol. 16(1), 2014, pp. 27-50.

[4] I-H. S. Kim, "Dynamic Management View: Logic of Profit Seeking Based on Adaptation to Technological Change and Needs Evolution Through Needs-Focused Innovation," Technology Analysis & Strategic Management, vol 29(10), 2017, pp. 1225–1242.

[5] B. Wirtz and P. Daiser, "Business Model Innovation: An Integrative Conceptual Framework," Journal of Business Models, vol. 5(1), 2017, pp. 14–34.

[6] I-H S. Kim, T-Y. D. Ku, and B-Y. M. Lee, Business model schema: business model innovation tool based on direct causal mechanisms of profit, Technology Analysis & Strategic Management, vol 32(4), 2020, pp. 379-396.

[7] O. Gassmann, K. Frankenberger, M. Csik, "Revolutionizing the Business Model," In: O. Gassmann and F. Schweitzer (Eds.), Management of the Fuzzy Front End of Innovation, Cham: Springer, 2014, pp. 89-98.

[8] B. Sinclar, IoT Inc., New York: Mc Graw Hill, 2017

[9] B. K. Panda, "Application of business model innovation for new enterprises. A case study of digital business using a freemium business model." Journal of Management Development, 39(4), 2020, pp. 517-524.

[10] A. Osterwalder, Y. Pigneur, and C. L. Tucci, "Clarifying Business Models: Origins, Present, and Future of the Concept," Communications of the Association for Information Systems 16 (1), 2005, pp. 1-25.

[11] H.W. Chesbrough, "Business model innovation: Opportunities and Barriers," Long Range Planning, vol. 43, 2010, pp. 354– 363.

[12] B. Spencer, "Business Model Design and Learning: A Strategic Guide," Business Expert Press, 2013.

Performance Analysis of Single and Multi-step Short-term Load Forecasts Using Multi-layer Perceptron

Athanasios Ioannis Arvanitidis
Electrical & Computer Engineering
University of Texas at San Antonio
San Antonio, USA
athanasios.arvanitidis@my.utsa.edu

Dimitrios Kontogiannis
Electrical & Computer Engineering
University of Thessaly
Volos, Greece
dimkonto@uth.gr

Georgios Vontzos
Electrical & Computer Engineering
University of Thessaly
Volos, Greece
gvontzos@uth.gr

Vasileios Laitsos
Electrical & Computer Engineering
University of Thessaly
Volos, Greece
vlaitsos@uth.gr

Dimitrios Bargiotas
Electrical & Computer Engineering
University of Thessaly
Volos, Greece
bargiotas@uth.gr

Miltiadis Alamaniotis
Electrical & Computer Engineering
University of Texas at San Antonio
San Antonio, USA
miltos.alamaniotis@utsa.edu

***Abstract*—Load forecasting is one of the most critical factors in modern power systems since it is the cornerstone for efficient monitoring, resource management and decision making. Therefore, there is an accrescent need for accurate and fast electrical load predictions. Many scientific approaches have been carried out in the field of load forecasting. In particular, the field of Machine Learning has attracted great research interest, due to the ability to adapt to time-series through forecasting tasks on multiple prediction horizons, a research area that presents challenges to several traditional methods. For this purpose, this research offers a thorough comparative study of several structural morphologies of Multi-Layer Perceptrons, in order to investigate electrical load forecasting accuracy for one, twelve and twenty-four time-steps ahead. Based on data from the Greek Power System for the years 2017 to 2019, the three proposed neural networks' structural morphologies are assessed in terms of precision through the Mean Absolute Error, Mean Squared Error, and Mean Absolute Percent Error of the predicted outcomes.**

***Index Terms*—multi-layer perceptron, univariate prediction, multivariate prediction, short-term load forecasting**

I. Introduction

Nowadays accurate load forecasting is crucial for power companies in order to sufficiently and reliably generate, transmit and distribute electric power. Non accurate load prediction could lead to limited facilities capacity, power supply shortage or even power interruption, and causes annoyance to stakeholders and consumers. Additionally, an accurate forecasting of electrical load supports proper infrastructures' maintenance and reduction of power companies' operational costs [1]. Load forecasting is commonly categorized as very short-term (VSLTF), short-term (SLTF), mid-term (MLTF) and long-term (LTLF) forecasting [2]. In this work, STLF is used to predict future demand.

In terms of electrical load forecasting several methodologies have been introduced and can be divided into two main categories the traditional and the modern methods. In traditional techniques, statistical methods are mainly utilized. These include models like Autoregression (AR), Moving Average (MA), Autoregressive Moving Average (ARMA), Autoregressive Integrated Moving Average (ARIMA), ARMA and ARIMA with exogenous inputs (ARMAX and ARIMAX respectively), Grey (GM) and Exponential Smoothing (ES) [3], [4].

On the other hand, modern load forecasting methods are considered machine learning, artificial intelligence-based and hybrid technics. Machine learning approaches include Support Vector Machine (SVM) models, which seems to be used extensively in forecasting issues. Additionally Artificial Neural Network (ANN) algorithms are very popular in recent years in time series prediction. The commonly used ANN algorithms for electrical load forecasting are Recurrent Neural Networks (RNN), such as Long-Short Term Memory (LSTM), Convolutional Neural Networks (CNN) and Feed-Forward Neural Networks, like Multi-Layer Perceptron (MLP) [5], [6].

Several recent studies propose the use of MLP models in future electrical demand prediction [7]. Although MLPs are the simplest type of ANN, they are used to complex non-linear problems. They not only perform well with large number of input data but also provide fast predictions after training [8], such as Kontogiannis et al. were observed at experiments with residential real world data compared with LSTM and CNN models [9]. Arvanitidis et al. used MLP models to propose novel train data pre-processing strategies [10] and clustering techniques [11] for SLTF. Furthermore, MLP architecture is extended to conduct a day-ahead electricity price forecasting [12]. MLP is impactful in short term forecasting tasks involving not only load related time series but also wind power forecasting. In [13], the Simulated Annealing optimization algorithm is used to specify the hyperparameters of the under investigation regressor models. The MLP model seems to predict wind power accurately compared to the other forecasting methods examined.

Electric load forecasting is a vital step in the planning of the power system industry. It is crucial to the management of the power system and the scheduling of electricity in order to ensure the system's cost-effective and uninterrupted performance. As a result, it offers multiple significant benefits for controlling generating capacity, scheduling, management, peak reduction, market assessment, and demand response. As a result, forecasting on diverse time horizons has shown to be incredibly effective in meeting the various criteria of their application. The literature provides a plethora of publications that investigate the subject of load forecasting in various prediction horizons, but without indicating which is the optimum and most efficient. Hence, the extensive comparison of univariate and multivariate short-term load forecasting methods is the novelty of this work. In this paper, three different optimized structural morphologies of MLPs are applied to anticipate load values for one, twelve, and twenty-four hours ahead, respectively. In order to determine which of the suggested morphologies of the neural networks delivers more accuracy, the results are compared based on the various prediction metrics.

The paper is organized as follows. Section II analyzes the necessary actions related to the preprocessing of the data used for forecasting, while Section III analyzes the forecasting model. Section IV describes the hyperparameters' optimisation algorithm, proposes three different MLP prediction models and presents the simulation results of short-term load forecasting, while Section V concludes the results of the paper.

II. Dataset Overview

The datasets utilized in this research project for the performance evaluation of the multilayer perceptron structure on total electricity load predictions for the Greek power system consist of hourly measurements of total load in MW, temperature in °C and relative humidity expressed as a percentage. The samples in this dataset cover a three year time period spanning from 2017-01-01 to 2019-12-31. The samples of total load were made publicly available by the transparency platform Entso-E [14] and the environmental variables of temperature and relative humidity were accessed through the MERRA-2 research and analysis platform [15].

Since this project considers the short-term forecasting tasks of 1, 12 and 24 hour-ahead total load prediction, the features utilized in this analysis include the temporal variables for the hour and day of the week encoded in the value intervals 0-23 and 1-7 respectively, the temperature and humidity variables for the target time interval and the historical load features corresponding to the same time intervals for the previous 7 days. The resulting datasets did not have any missing values and the values were scaled through min-max normalization [16]. The dataset was split into a training set containing 80% of samples and a test set containing 20% of samples based on common practices with regards to the splitting ratio [17]. Figure 1 presents the target variable of load.

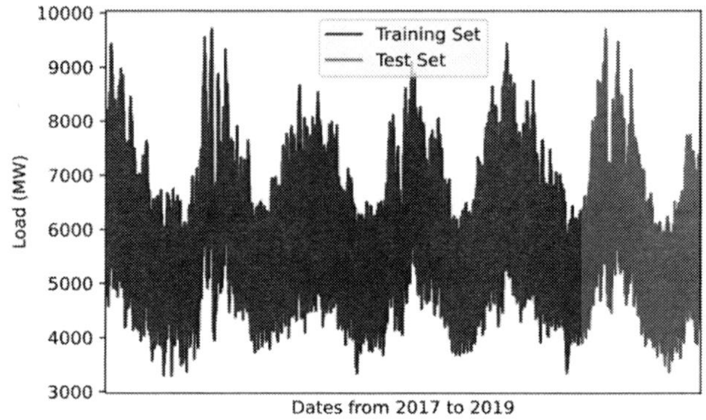

Fig. 1. Graphical representation of the dataset load values.

III. Multi-Layer Perceptron

The MLP is a neural network structure that belongs to the class of feed forward artificial neural network architectures as it expresses a fully connected acyclic computation graph that focuses on the task of function approximation in order to derive a model that efficiently predicts the target variables given the input features x through the target function that describes the relationship between the inputs and the output, defined as f. This neural network consists of neurons that perform the computations following the perceptron supervised learning algorithm [18]. These neurons are organized in layers that express different roles in the computation path. The input layer receives the features from the initial input dataset, the hidden layers express the transition from the input to the output through a series of computations involving the adjustment of a weight matrix W that quantifies the importance of the input features towards the prediction of the target output and a bias vector b that is used to offset the computed results of the neurons. The output of each hidden layer is determined based on activation functions that evaluate the importance of computations and select the data that will proceed deeper in the network. The output layer derives the estimated values from the output of the last hidden layer. Consequently, the estimated values $f(x)$ derived from a multi-layer perceptron with one hidden layer could be expressed by formula (1) denoting the subsequent adjustments to weight matrix W^1 and bias vector b^1 for the transition from the input to the first hidden layer through the activation function as well as the application of the activation function and the adjustments to the weight matrix W^2 and bias vector b^2 for the transition from the hidden layer to the output layer.

$$f(x) = G(b^2 + W^2(s(b^1 + W^1x))) \tag{1}$$

Since this study examines univariate and multivariate load forecasting tasks, it is worth noting that the MLP is modified accordingly in order to derive the appropriate estimated values. Therefore, the number of neurons in the output layer needs to match the number of predicted output variables, corresponding to the target time series. The neural networks with a single output neuron are known as univariate MLPs and are used for hourly load forecasting, while the MLPs with several neurons in the output layer are known as multivariate and are employed in day ahead load prediction, as Figure 2 depicts. The MLP structure is typically trained through back propagation with gradient descent. Impactful structural parameters that could affect the training process and the performance of the model include the number of neurons at each layer, the number of hidden layers and the number of training epochs. Additionally, learning parameters, such as learning rate, the types of activation functions and the optimizer could be equally important to the generalization capabilities of the model [19].

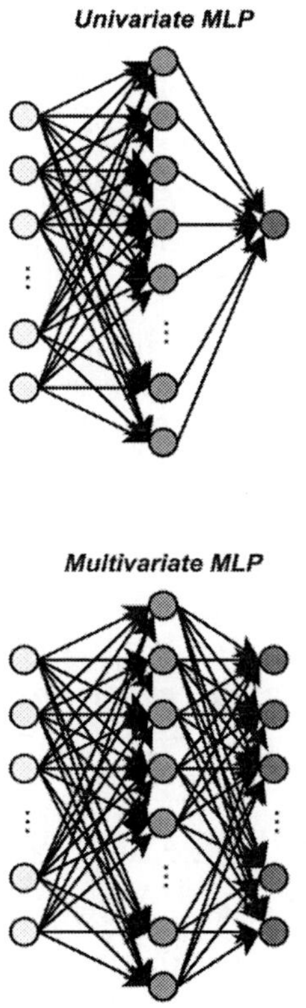

Fig. 2. Structural difference between univariate and multiariate MLPs.

IV. Evaluation Metrics

In this section, the metrics used for the performance evaluation of the single and multi-step forecasting models are outlined in order to further explain their role in our experiments. Firstly, Mean Absolute Error (MAE) is utilized as a simple and interpretable metric in order to naturally describe the average error of the MLP. Secondly, Mean Squared Error (MSE) is included in the examination of model performance since it is a scale dependent error metric that considers the direction of the predicted values. Lastly, Mean Absolute Percentage Error (MAPE) was utilized in order to denote the generalized relative error of the models. Given the forecasted data points y_i and the actual values x_i in a set of n observations, MAE, MSE and MAPE are given by (2), (3) and (4) respectively [20].

$$MAE = \frac{\sum_{i=1}^{n} |y_i - x_i|}{n} \tag{2}$$

$$MSE = \frac{1}{n}\sum_{i=1}^{n}(y_i - x_i)^2 \tag{3}$$

$$MAPE = \frac{100}{n}\sum_{i=1}^{n}|\frac{y_i - x_i}{x_i}| \tag{4}$$

V. Results

This study investigates the implications of MLPs' univariate and multivariate prediction procedures on the precision of short-term load forecasting. Also, the variations in the execution time of a brute force optimization algorithm, which is employed for the optimal hyperparameter selection on the various morphologies of the proposed MLPs, are thoroughly addressed. The suggested MLPs were developed using Python's scikit-learn framework [21], while the computer system utilized has an Intel Core i7-4510U CPU running at 2.00 GHz and 8 GB of installed memory.

Initially, the suggested brute force optimization approach is used to determine the ideal values for two major neural network hyperparameters: the number of neurons in the hidden layer and the number of epochs throughout the training process. For the sake of simplicity, each suggested MLP comprises of a hidden layer, as it is adequate for the load forecasting issue. Most neural networks used for load forecasting include only one hidden layer in order to reduce computational complexity while yet providing fast online results.

In this paper, we examine the structural morphology of three neural networks in order to estimate hourly load values for one hour, twelve hours, and twenty-four hours ahead, respectively. Thus, in pursuance of a direct comparison on the hyperparameter selection, we suggest that the number of neurons in the hidden layer is determined as a function of the number of neurons in the input layer for each possible structural morphology. In each case examined, the minimum acceptable number of hidden neurons is half of the number of neurons in the input layer, while the maximum number of hidden neurons can reach up to three times the number of input neurons. The ideal number of epochs, on the other hand, is derived by sequential scanning within the closed interval [200, 2000], where 200 is the least permitted limit of iterations and 2000 is the maximum limit of epochs.

In the following subsections, the calculation of the hyperparameters of the three different proposed MLPs, as well as the execution time of the brute force optimization approach for each case and their performance evaluation, in terms of accuracy, for short-term load forecasting are thoroughly examined.

A. One hour ahead load forecasting - Univariate Neural Network

The case of a univariate MLP, i.e., a neural network with a single neuron in the output layer, is first investigated. This network generates the hourly value of the load for the following hour. The input data to this MLP, which consists of 11 neurons in the input layer, is as follows:

- A label for the time for which the forecast is being performed, represented as an integer within [0, 23].
- An integer that belongs to the range [1, 7], serving as a label to identify the day being predicted. Sunday is represented by the value 1, Monday by the value 2, etc.
- The hourly temperature value for the precise time of day for which the prediction is conducted.
- The hourly humidity estimation for that particular hour of day in which the prediction is made.
- Seven hourly load values for the period from the current time up to one week in beforehand of the prediction.

B. Twelve hours ahead load forecasting - Multivariate Neural Network

The case of an MLP used for 12 hours ahead load forecasting is then considered, i.e. a neural network with 12 neurons in the output layer. In this scenario, the number of neurons in the input layer is 109 and results from the following input data:

- An integer within the range [1, 7], serving as a label to identify the day being predicted. Sunday is represented by the value 1, Monday by the value 2, etc.
- A vector consisting of 12-hourly temperature values for specific hours of the day for which the prediction is conducted. A day is divided into two instances in the dataset. Thus, the first vector of the day concerns the hours from midnight to 11 am and contains the corresponding hourly temperature data.
- A vector consisted of 12-hourly humidity estimations, similar to the case of temperature, for that particular hour of day in which the prediction is made.
- A vector of 84-hourly load values concerning the period from the current time up to one week in beforehand of the prediction, respectively.

C. Twenty-four hours ahead load forecasting - Multivariate Neural Network

The last case study focuses at a multivariate MLP that is used to estimate day-ahead load and has 24 neurons in the output layer, one for each hour of the day. Similar to the other examples, the number of input data determines the quantity of neurons in the input layer. Thus, in this case, the 217 input neurons result from the following data:

- An integer in the range [1, 7], acting, as in the earlier cases, as a label to designate the day being forecast.
- A vector consisting of 24-hourly temperature values for the day of which the prediction is conducted.
- A vector consisted of 24-hourly humidity values for the day of which the prediction is conducted.
- A vector of 168-hourly load values concerning the period from the current time up to one week in beforehand of the prediction, respectively.

The boundaries of the hyperparameters for each MLP utilized in each case study are reported in Table I. The results of the optimization process used to identify the ideal hyperparameters for the MLPs of each case study are summarized in Table II. Subsequently, the optimized neural networks are used for load prediction and their results are compared, in terms of accuracy, in order to decide whether the univariate or the multivariate structural morphology responds better to the STLF issue. Furthermore, Table III compiles the findings of the MAE, MSE, and MAPE metrics yielded from each case study. Lastly, Figure 3 and Figure 4 graphically represent the STLF outcomes for each prediction method considered.

TABLE I
RESULTS OF THE OPTIMIZATION APPROACH FOR EACH CASE STUDY.

	Boundaries		Step	
MLP	*Neurons*	*Iterations*	*Neurons*	*Iterations*
1h Ahead	[6, 33]	[200, 2000]	6	10
12h Ahead	[55, 327]	[200, 2000]	55	10
24h Ahead	[109, 651]	[200, 2000]	109	10

TABLE II
RESULTS OF THE OPTIMIZATION APPROACH FOR EACH CASE STUDY.

MLP	*Neurons*	*Iterations*	*Time (H:MM:SS)*
1h Ahead	33	2000	2:20:11
12h Ahead	275	1800	0:56:59
24h Ahead	436	1800	1:19:43

TABLE III
ACCURACY METRICS DERIVED FROM EACH MLP FOR STLF.

MLP	*MAE*	*MSE*	*MAPE (%)*
1h Ahead	182.076	67603.22	2.774
12h Ahead	162.845	54383.46	2.435
24h Ahead	187.315	66564.93	2.742

VI. CONCLUSION

In this paper, a detailed comparative analysis was conducted with the purpose of providing accurate load forecasting results for the Greek Power System for the period 2017-2019. The implementation of three different structural morphologies of MLP models were developed and assessed.

Based on the results, it is concluded that load forecasting with the implementation of MLPs plays a critical role in the safety, stability, and sustainability of modern energy systems. More specifically, it is observed that for all three cases, the results are quite satisfactory. The error values for the one hour ahead, and twenty four hours ahead forecast are very similar in terms of error metrics. The twelve hour ahead model exhibited improved performance compared to the other forecasting horizons. This might be because the number of

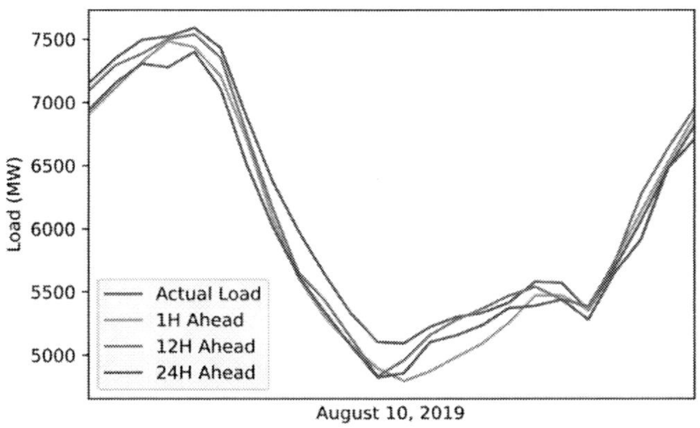

Fig. 3. Graphical comparison of the load prediction results of each considered method for a whole day in August.

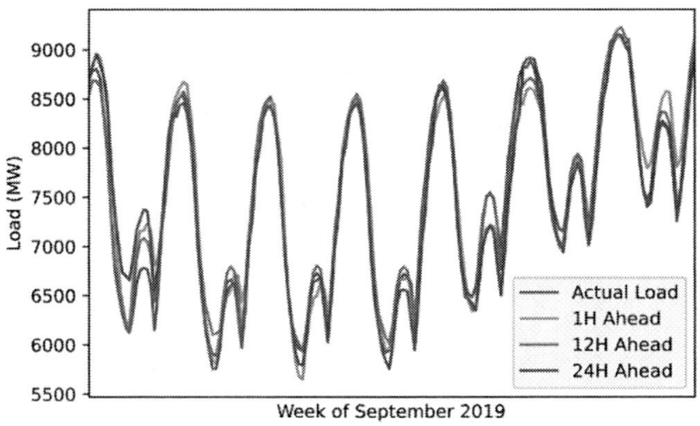

Fig. 4. Graphical comparison of the load forecasting results of each considered method, for a week in September.

input layer neurons, and hence the quantity of data utilized as input data, more closely approximates the case study under discussion. Given the data quality and data seasonality of the dataset, all models yielded relatively low error metrics. This observation proves the correct use of the algorithm in a time-series with multiple seasonalities, like the one studied. Since, the most accurate forecast was found for the case of twelve hours ahead, the ability of the algorithm to adapt to multi-step ahead forecasting is highlighted. Finally, it is worth noting that these models could assist the uninterrupted and reliable operation of Smart Grids using real-time data, where day ahead load forecasting delivers significant value.

In future work, this model could be evaluated on more complex load forecasting issues and compared with other deep learning models, used as benchmarks. Also, the proposed techniques could be applied to Demand Side Management and Demand Response programs [22], which have developed rapidly in recent years due to the global increase of energy consumption.

References

[1] E. A. Feinberg and D. Genethliou, "Load forecasting," in *Applied mathematics for restructured electric power systems*. Springer, 2005, pp. 269–285.

[2] A. A. Mamun, M. Sohel, N. Mohammad, M. S. Haque Sunny, D. R. Dipta, and E. Hossain, "A comprehensive review of the load forecasting techniques using single and hybrid predictive models," *IEEE Access*, vol. 8, pp. 134 911–134 939, 2020.

[3] M. A. Hammad, B. Jereb, B. Rosi, and D. Dragan, "Methods and models for electric load forecasting: a comprehensive review," *Logistics, Supply Chain, Sustainability and Global Challenges*, vol. 11, no. 1, pp. 51–76, 2020.

[4] T. Hong and S. Fan, "Probabilistic electric load forecasting: A tutorial review," *International Journal of Forecasting*, vol. 32, no. 3, pp. 914–938, 2016.

[5] I. K. Nti, M. Teimeh, O. Nyarko-Boateng, and A. F. Adekoya, "Electricity load forecasting: A systematic review," *Journal of Electrical Systems and Information Technology*, vol. 7, no. 1, pp. 1–19, 2020.

[6] A. Román-Portabales, M. López-Nores, and J. J. Pazos-Arias, "Systematic review of electricity demand forecast using ann-based machine learning algorithms," *Sensors*, vol. 21, no. 13, p. 4544, 2021.

[7] A. Gasparin, S. Lukovic, and C. Alippi, "Deep learning for time series forecasting: The electric load case," *CAAI Transactions on Intelligence Technology*, vol. 7, no. 1, pp. 1–25, 2022.

[8] "Deep Learning Tutorial for Beginners: Neural Network Basics." [Online]. Available: https://www.guru99.com/deep-learning-tutorial.html#5

[9] D. Kontogiannis, D. Bargiotas, and A. Daskalopulu, "Minutely active power forecasting models using neural networks," *Sustainability*, vol. 12, no. 8, p. 3177, 2020.

[10] A. I. Arvanitidis, D. Bargiotas, A. Daskalopulu, V. M. Laitsos, and L. H. Tsoukalas, "Enhanced short-term load forecasting using artificial neural networks," *Energies*, vol. 14, no. 22, p. 7788, 2021.

[11] A. I. Arvanitidis, D. Bargiotas, A. Daskalopulu, D. Kontogiannis, I. P. Panapakidis, and L. H. Tsoukalas, "Clustering informed mlp models for fast and accurate short-term load forecasting," *Energies*, vol. 15, no. 4, p. 1295, 2022.

[12] D. Kontogiannis, D. Bargiotas, A. Daskalopulu, A. I. Arvanitidis, and L. H. Tsoukalas, "Error compensation enhanced day-ahead electricity price forecasting," *Energies*, vol. 15, no. 4, p. 1466, 2022.

[13] A. I. Arvanitidis, D. Kontogiannis, G. Vontzos, V. Laitsos, and D. Bargiotas, "Stochastic heuristic optimization of machine learning estimators for short-term wind power forecasting," in *2022 57th International Universities Power Engineering Conference (UPEC)*. IEEE, 2022, pp. 1–6.

[14] "Central collection and publication of electricity generation, transportation and consumption data and information for the pan-european market." [Online]. Available: https://transparency.entsoe.eu/load-domain/r2/totalLoadR2/show?name=

[15] "Merra-2." [Online]. Available: https://gmao.gsfc.nasa.gov/reanalysis/MERRA-2/

[16] S. K. Patro and K. K. sahu, "Normalization: A preprocessing stage," *IARJSET*, p. 20–22, 2015.

[17] "Split training and testing data sets in python," Oct 2020. [Online]. Available: https://www.askpython.com/python/examples/split-data-training-and-testing-set

[18] "Dive into deep learning." [Online]. Available: http://d2l.ai/chapter multilayerperceptrons/mlp.html

[19] [Online]. Available: https://towardsdatascience.com/what-are-hyperparameters-and-how-to-tune-the-hyperparameters-in-a-deep-neural-network-d0604917584a

[20] "Error metrics: How to evaluate your forecasting models," May 2022. [Online]. Available: https://www.jedox.com/en/blog/error-metrics-how-to-evaluate-forecasts/

[21] F. Pedregosa, G. Varoquaux, A. Gramfort, V. Michel, B. Thirion, O. Grisel, M. Blondel, P. Prettenhofer, R. Weiss, V. Dubourg, J. Vanderplas, D. Passos, A. and Cournapeau, M. Brucher, M. Perrot, and E. Duchesnay, "Scikit-learn: Machine learning in Python," *Journal of Machine Learning Research*, vol. 12, pp. 2825–2830, 2011.

[22] V. M. Laitsos, D. Bargiotas, A. Daskalopulu, A. I. Arvanitidis, and L. H. Tsoukalas, "An incentive-based implementation of demand side management in power systems," *Energies*, vol. 14, no. 23, p. 7994, 2021.